T0219987

Prototyping Python Dashboards for Scientists and Engineers

Build and Deploy a Complete Dashboard with Python

Padraig Houlahan

Apress®

Prototyping Python Dashboards for Scientists and Engineers: Build and Deploy a Complete Dashboard with Python

Padraig Houlahan
Flagstaff, AZ, USA

ISBN-13 (pbk): 979-8-8688-0220-1   ISBN-13 (electronic): 979-8-8688-0221-8
https://doi.org/10.1007/979-8-8688-0221-8

Copyright © 2024 by Padraig Houlahan

Managing Director, Apress Media LLC: Welmoed Spahr
Acquisitions Editor: Celestin Suresh John
Development Editor: James Markham
Editorial Assistant: Gryffin Winkler

Cover designed by eStudioCalamar

Cover image designed by Freepik (www.freepik.com)

Distributed to the book trade worldwide by Springer Science+Business Media New York, 1 New York Plaza, Suite 4600, New York, NY 10004-1562, USA. Phone 1-800-SPRINGER, fax (201) 348-4505, e-mail orders-ny@springer-sbm.com, or visit www.springeronline.com. Apress Media, LLC is a California LLC and the sole member (owner) is Springer Science + Business Media Finance Inc (SSBM Finance Inc). SSBM Finance Inc is a **Delaware** corporation.

For information on translations, please e-mail booktranslations@springernature.com; for reprint, paperback, or audio rights, please e-mail bookpermissions@springernature.com.

Apress titles may be purchased in bulk for academic, corporate, or promotional use. eBook versions and licenses are also available for most titles. For more information, reference our Print and eBook Bulk Sales web page at http://www.apress.com/bulk-sales.

Any source code or other supplementary material referenced by the author in this book is available to readers on GitHub. For more detailed information, please visit https://www.apress.com/gp/services/source-code.

Paper in this product is recyclable

To my mother, Kathleen (Moriarty) Houlahan. Teacher.

(Born 1927, Brandon, County Kerry, Ireland)

Table of Contents

vii

About the Author

Padraig Houlahan has a diverse career, spanning research, teaching, and IT management. He has worn multiple hats over the years, functioning as a scientist, software developer, college professor, flight instructor, and IT director. Throughout his journey, he has maintained an enduring fascination with the art of uncovering patterns within data. His M.A. research involved searching for, and placing upper limits on, gravitational waves in Lunar Laser Ranging data, while his Ph.D. research introduced an innovative pattern recognition approach for characterizing the structure of astronomical nebulae. Since then, his focus has pivoted toward aviation-centric software projects. These ventures involve real-time tracking and visualization of aircraft to aid in traffic flow analysis. Furthermore, he has played a pivotal role in rendering extensive datasets accessible to both students and faculty, benefiting airline and airport operators alike. Teaching has always held a special place in his heart, driven by the joy of witnessing students derive satisfaction from acquiring new knowledge and insights.

About the Technical Reviewer

 **Shibsankar** has 10+ years of experience working in IT where he has led several data science initiatives, and in 2019, he was recognized as one of the top 40 data scientists in India. His core strength is in deep learning, NLP, and graph neural networks, whereas he has also spent time solving problems in computer vision. He has experience working in the domain of foundational research, fintech, and ecommerce.

He is currently working at Optum as a principal data scientist. Before that, he worked at Walmart, Envestnet, Microsoft Research, and Capgemini. He pursued a master's from the Indian Institute of Technology, Bangalore.

Acknowledgments

While writing this book was a solitary endeavor, it could not have happened without the influence of an ecosystem of family and friends who help me keep my sanity, an ecosystem that includes the following people.

My wife Tania, a voracious reader and lover of animals and of the outdoors, who faces the daily challenge of managing her "grumpy old man in training" all the while working in physical therapy, teaching Aikido, learning mandolin, and wildlife sketching.

My mother, a remarkable individual born in a very remote part of Ireland, one of the most intellectually curious people I ever met, who became a teacher and who cultivated a love of learning and culture in all her children.

My friend Dr. Otto Franz, an astronomer whom I met while working at Lowell Observatory, now retired and living in Texas, who still stays in contact with weekly Facetime sessions, with whom I have the privilege of discussing topical and scientific issues with someone who was a child in Austria during the Nazi occupation, who offers eyewitness testimony and insights into extremism, historic and current. His appreciation of the importance of bringing data to life, and seeing the merits in my work, is greatly appreciated.

My friends Bill Burke (a superb luthier and musician) and his wife Pat (who meticulously documents historic archeological sites) for their regular wonderful conversations – on all things; and close friends Tim Sober and Carol Kennedy, Barry Malpas and his wife Anne Wittke, and Klaus Brasch for our regular conversations and sanity-saving rambling social and political reviews.

ACKNOWLEDGMENTS

And I must also mention Dr. Darrel Smith, a long-time friend and professor of Physics, Dr. Jules Yimga (Chair, School of Business), and Dr. Zafer Hatahet (Dean, College of Arts and Sciences) all at Embry-Riddle Aeronautical University (Prescott, AZ) for their positive responses to the software development efforts which are the focus of this book.

Truly, it does take a village to raise a child!

Introduction

I wrote this book for a past me. I'm now retired but have maintained a lifelong interest in programming; there's just something very satisfying in developing a piece of code that solves a problem. Frankly, it's empowering; you have a solution that gets the job done. I have had the good fortune of having worked on software projects that were done mainly to support academic research – research in astronomy and aviation-related studies, which have been my main interests – not programming. I like programming, but it's simply a way to get me to where I want to be. Sometimes, it's the journey, but for me, it's mainly the destination.

I am not alone. There are countless researchers and grad students and data analysts who program, but whose main interest is not programming. Being able to code is simply a necessary skill. It's the norm for academic types to write unique custom software tailored to a specific purpose or a narrowly crafted dataset. University departments, whether scientific or business or engineering, will rarely devote time to teaching graduate student workers and young faculty to program; they are expected to sink or swim. They will write code that would fail to consider multitudes of possible scenarios or input format, for example. "Need to enter a number? Well, make sure you use three significant digits for that column, or the code might crash!" – this probably happens more often than many would like to admit. The point here is not to assign blame, but to be realistic; very often, code is written in a very functional, get-it-done style, to simply... get it done. It's not about being the cleverest or the fastest, but simply getting to the destination.

There will be conflicting forces. As a team member, there is often a need to share your code and make it available for others to use, so they can get their work done. An Object-Oriented Design makes this much easier. Sometimes, the code takes on a life of its own and becomes a badly needed resource, a service if you will, to a broader interest group. Your supervisor or manager is not really interested in the details, just the results, and so a little more effort on the journey is warranted. And this is where dashboards enter the picture – they represent an encapsulation of your knowledge codified into a controlled look and feel and access. A particular kind of programming really, but to the nonspecialist, another hill to climb.

There have been many times when I found myself in such a position and wished I had some complete examples to look at, to see how things were done, and so this is the sense in which I say this book is written to a past me. In showing how a dashboard was built around an important aviation dataset, I hope I can help others benefit and expedite their own projects without facing the often-daunting headwinds of learning new technologies and techniques all by themselves.

Why Dashboards?

A picture is worth a thousand words.

—Fred R. Barnard

All data is ultimately a collection of bits – ones and zeros really. It can be stored in various forms from disks to paper, but to be useful it needs to be understood and accessible, both of which benefit from being able to readily visualize what the data represents and from being able to analyze and share it. Dashboards (graphical displays using features like buttons, sliders, menus, and graphics) have emerged as powerful tools to address these needs. When well designed, you can encapsulate the problems of data management, display, and access into a self-contained project, allowing collaborators to work remotely if needed, all the while providing a focus for the research effort's needs to manage the data centrally.

Well-designed dashboards should make their data as intuitive and functional to the user as possible. By intuitive, I ideally mean the user should see compelling graphics and results that are easy to comprehend. It might not be possible to avoid long tables of data, but in general, well-designed graphical elements are more effective than raw numbers for initial explorations. Part of this is simply human nature. We are visual creatures. Of course, students and researchers need to get into the weeds with the algorithms and raw data, but eventually, that data must be reduced into usable summaries and statistics for practical use by themselves and external users.

To be functional, a dashboard needs to meet the designers' goal of making the data's attributes of interest accessible. A large dataset might contain more information than the designer is interested in. So, the designer's task is to figure out what part of the dataset to use and to create displays and tools to reveal the embedded traits of interest; the designer must regularly choose between too little and too much. If they include too much information, they run the risk of overwhelming the nonexpert; too little, and their dashboard appears to be inconsequential. For these reasons, designers must always keep the end user in mind and be willing to revise their dashboards accordingly.

There are multiple ways a dashboard might be built and accessed like web based, application based, and desktop based. A web-based dashboard has an advantage as it offers freedom to the developer to maintain an operating system–specific environment and configurations. This makes them more stable and greatly simplifies the development process – the designer only needs to develop for a single platform. At the same time, the distribution problem is solved. A dashboard on a server can be shared with any other computer on the Internet.

Good dashboard projects are therefore a reflection of carefully planned data analysis, reduction, presentation, management, and dissemination.

The Importance of Data Visualization

The main dashboard we will build in this book allows the user to explore the government's airport operations dataset (ATADS) that tracks various types of daily operations for more than 500 US airports. Operations include the kind of flights (military, civilian, commercial), the conditions (whether they were conducted using Instrument Flight Rules [IFR] or Visual Flight Rules [VFR]), and whether they were local or itinerant. VFR operations are permitted when the weather conditions meet certain minimum visibility and cloud clearance criteria and are normally used by private pilots and small air taxi or commercial companies. One cannot assume IFR flights happen under inclement weather since larger commercial airlines will use these by default for safety.

A dataset such as this contains a wealth of hidden information that is hard to appreciate without seeing it graphically, so we can answer basic questions concerning activity trends. But if we pay close attention, we will also discover other features that would be difficult to detect from inspection of tabular data by itself, some of them being quite fascinating in their own right, and so a graphical inspection of the data is very important since it will very likely lead to further questions to explore.

This ability to support visual inspection of graphical data is an important feature of many dashboards and is essential in ours. After we are done developing and deploying our dashboard, in a later chapter, we will demonstrate how powerful it is in providing insight into the ATADS dataset and hopefully encourage and motivate you to explore your project's data.

Finally, an important aspect of this book is not just showing how to design, build, and deploy a dashboard, but since we are using an important dataset from the aviation industry, we also address issues of how our dashboard can support research and aviation professionals and business students pursuing a career in aviation management. There are scenarios we will explore as part of demonstrating the project's usefulness, and it is hoped that all readers will appreciate how what might be perceived as a fairly dry dataset actually presents wonderful opportunities for, and naturally suggests, topics for further exploration. Having been a researcher and scientist, I have always enjoyed digging deep into data to see what's hidden beneath the surface, which is why I think it's important to explore how the dashboard can help reveal and document behaviors and events such as hurricanes and pandemics. Hopefully, these explorations will motivate those interested in other datasets.

Why This Book?

This book is intended for those who work with data and also have the need to share insights regarding data. An essential aspect is a desire to help their end user "lift the veil" and understand underlying data features such as trends and patterns. Large organizations can have teams of developers and experts to achieve these goals, but many who could benefit from such capability (e.g., college faculty and graduate students) work with minimal resources and need to take a jack-of-all-trades approach to building applications; they need to solve the data access, data importation, data display, analytical tool build, server operating system management, and deployment problems (if your dashboard is wildly successful, you will need a scalable solution!). All these must be done in a realistic manner. Some corners can be cut – there are fewer deployment concerns if the project is behind a firewall or if the users are all experts, and primitive graphics can work if there are deadline constraints. Regardless, there is a daunting task and skillset list the graduate or faculty member must master.

This book is intended to show how a complete dashboard-based project can be created and deployed. I will use some government aviation datasets as core material, and I will show the various design issues and solutions I encountered in developing an online dashboard. In taking a step-by-step approach, I hope newcomers will see how this can be done and encourage them to do it for themselves. I hope the reader finds it useful to have the diverse underlying issues addressed in one volume with personal observations included along the way.

There are many online resources that can show you how to make a simple dashboard or that offer solutions created by others that can be licensed. There are few, if any, that demonstrate a complete solution incorporating real-world data access and download, comprehensive reactive programming, and server configuration issues. The obvious reason is that it takes a considerable effort to organize and present such a complete overview, such as this work, and this is beyond the scope of most online articles. Perhaps not too unfairly, many online examples teach how to draw a stickman when what you really need to know is how to do a portrait.

In this book, we address the "complete solution" for the core application and provide the user with complete code examples and the needed server configurations. This should serve as a powerful self-contained resource to learn from, and to refer to, since there were many times when working on various research projects, I found myself at a point where I felt I understood the broad strokes of a coding issue, but, frustratingly, just needed a little extra help in the form of seeing how someone else solved an issue, and this was all I needed to move forward. No question having complete examples to refer to is both a tremendous time saver and stress reliever.

There is no need to be a computer expert to achieve this, but computer expertise is still needed. Most researchers and students these days have experience with some programming, system administration, and basic web HTML hacking and could learn how to build dashboards for themselves. For this reason, when discussing code examples, only the major features will be addressed; the assumption will be that the reader will be able to understand a function's working from reading the code combined with their understanding of coding basics.

I have also taken the liberty of addressing some issues related to data analysis and interpretation for aviation professionals specifically interested in aviation datasets, and this goes beyond the basic task of creating a dashboard. I include a discussion showing how our dashboards can provide insight into many aviation issues ranging from the impact of hurricanes to pandemics, and I even include discussions on how to create models to test your understanding and to help incorporate spectral analysis into your work. I realize spectral analysis of airport operations data is surprising perhaps, but why not? The techniques are universal, and it was interesting to see how it worked out and it raised interesting questions in its own right. And, of course, these discussions can be generalized to non-aviation projects. So, in this book, we will build a custom tool and show how to use it, which is a valuable part of the feedback used to refine our designs.

I will note that dashboards will evolve. In other words, you don't have to achieve every desirable outcome in your first attempt. Start with a simple goal such as displaying a time series of a single data type. Figure out how to import and display the data; at this point, you have made significant inroads into the I/O and graphics display problem. Next, start adding tools like linear models for trend analysis or panels to display important statistical values. And that's it! Well, almost. As you work with your data and your dashboard, you probably will have "Aha!" moments where new ways to visualize the data pop into your mind, and so displays will have to be modified or new tools created. The project will evolve. However, even as large datasets are continuously updated, their corresponding dashboards should eventually stabilize. Done right, this is a wonderful outcome for any research project, a resource to be shared with end users – public or academic. And let's not forget the benefits your colleagues will gain by not having to reinvent the wheel.

How to Use This Book

Depending on what programming challenges you face and your stage of completion, your approach to how you might use this book will vary greatly. It would fall into one or more of the following strategies:

- As a general reference on how to build and deploy a Python dashboard.

- As a template: Here, you would download the code examples from the publisher's website to explore the various solutions and to modify them to suit your purpose. You would provide your own CSV file collection and modify the names and labels so that the drop-down menus and equation strings would be appropriately populated. The result would be a dashboard tailored to your project.

- As a resource: See how to work with PLOTLY/DASH when first learning reactive programming.

- As a guide on how to deploy a Python application using NGINX and GUNICORN, with a discussion of how to use CSS to control your dashboard's screen layout.

- As a resource to augment your current development efforts: For example, to see how borders, titles, equation strings, and mouseover text are created and curve fit and Fourier techniques implemented.

- As a how-to when getting started with some file format translation (e.g., when converting XLS to CSV) and in screen scraping (i.e., automatically navigate a complex website to customize data for download).

- As an introduction on how to construct a WordPress web portal and on how to track down arcane formatting issues using Chrome Developer Tools.

No matter which strategy is of interest to you on a particular day, my goal is to help you see how a functioning dashboard is constructed and deployed by giving you access to the underlying code and server configuration details, for you to learn from and leverage the learning while developing a dashboard, so you can break through whatever roadblock is distracting you from your project's objectives.

Bits and Pieces

The projects I describe in this book use Python and PLOTLY/DASH for code development.

DASH is a free, open source software available under the MIT license, and the license file which contains the copyright statement can be found at https://github.com/plotly/dash/blob/dev/LICENSE. (Same for Plotly.py – https://github.com/plotly/plotly.py/blob/master/LICENSE.txt.)

I use the Anaconda code management environment to access the Spyder Python development environment. I find this combination works well for me. Spyder can be run by itself, but running it under Anaconda has generally made software updates more seamless. Spyder is free and lightweight. It's effective for me. It has limitations and sometimes I've been frustrated by it, but it delivered. There are other flavors of Python – feel free to use whatever works for you. Ultimately, you will have code in a ".py" file – it matters not whether it came from Spyder or PyCharm or whatever.

While running a Plotly application in the Spyder IDE, it starts an application server. By default, the dashboard is accessible via a browser at http://127.0.0.1:8050 where 127.0.0.1 is the IP address of the localhost server of the local computer and 8050 is the default port for Plotly/DASH. As we will discuss later, when we deploy our dashboard, we will need to offer different port numbers to different users, so that each user can experience a personalized view.

To handle the complex problems of displaying results on a screen, such as with a dashboard, we do not have to reinvent the wheel, we can use software libraries created by experts who specialize in this problem. Again, there are many possibilities, but I work with PLOTLY – a commercial product that faculty and students can use for free. If your project has commercial implications, you will of course need to talk to them. I personally like the PLOTLY solutions because I find they are well documented for the most part, and I think their support forums are very effective – they seem to go to great efforts to help their users. And, while not perfect, they provide some really good foundational code examples demonstrating their very comprehensive capabilities.

In some ways, creating a new dashboard is like trying to find your way through a swamp of data and technology issues to get to the other side. I just show the path I took that worked for me. There are many others, but this approach is one that worked and is well suited to small teams or individual developers.

Beyond the basic software coding, there are some other major configuration efforts: mainly setting up the web server and the system configuration. I used NGINX because it is a widely used lightweight server, and then the operating system needed appropriate servers configured to service NGINX, the dashboard. There are some other tasks I'll cover, but all this complexity can be handled if broken down into appropriate steps.

Prototyping

This is a book intended to be a resource for professionals, researchers, and students, whether they are involved in university projects or airline IT departments. Airlines, airport operators, researchers, and students might want to develop customized dashboards to support their specific needs. I emphasize time series data, daily records of activity, from which trends and patterns might be gleaned to help with forecasting. Obviously, any CSV time series data could be used – shipping traffic, hotel occupancy, weather stations, etc. Our use of one industry's dataset helps make our challenges more focused and less abstract.

When prototyping, functionality will be a core design philosophy, and at the early stages of a project, there will be many cycles where designs are abandoned or changed daily; documentation might be sparse, and the code should be as self-documenting and intuitive as possible. Sometimes, brute-force, inelegant solutions will be applied to simply get the job done, with a promise to oneself to clean things up or to automate a chore later. Object-Oriented Design (OOD) objects will become bloated until they are intolerable to work with, and refactoring (code redesign) will be done to restore clarity. There will also likely be dead pieces of code left in a file, commented out, but kept there because it worked or had a clever algorithm that might still be useful. This is the real world of prototyping.

When developing complex coding solutions, it is essential to understand the technical stuff, but there is a reason why websites offer step-by-step solutions on how, for example, to install and configure a MySQL server; having access to solutions is an extremely efficient way to get things done, to bypass a possibly considerable burden in time and effort required to build something from scratch. This is one of this book's main goals – to allow the reader to see how a complex dashboard was implemented. Some code segments will look ugly; some solutions need to be better automated; some OOD objects should be rewritten. And yet, what a time saver it would have been for me if I knew then what I know now and picked up this book – warts and all – to help me get going!

Being Organized – Managing Your Project

Whether you're embarking on a new project expected to occupy your focus in the foreseeable future or already knee-deep in prototyping, effective effort management is crucial. It's essential to initiate this process promptly if you haven't done so already.

Keeping notes in a manila folder is OK, but I strongly recommend you buy a large hardcover notebook such as one found in a college or office supply store. I like to use a black pen – the contrast it provides is easier on your eyes. Let your notes breathe – writing notes in the equivalent of an 8pt font or with a pencil will make your notes too hard to reread; leave plenty of space between the lines. Number your pages and leave four or five pages at the beginning of the notebook, so you can build a table of contents on the fly with page number, date, and a few words describing that page's contents. With this, you will be able to find work done months ago with minimal effort. I would suggest writing comments and keywords at the top of each page or day's work, giving a little more information than what you entered in your table of contents. I keep critical information like server IPs, names, and accounts at the back of the notebook, so I can quickly find them as needed. Critical code segments can be printed out and taped/ glued into the notebook. When glancing at an older entry, you can always add a quick note to see a later page for clarity. Yes, all of this is old school, but it will give you a sense of accomplishment and progress and will help keep concerns your project is becoming overwhelming at bay since you will be able to quickly reorient yourself with forgotten ideas and solutions.

Your notebook, if done properly, will document your prototyping. It's a balance between how much or how little to include. Too much becomes a chore, and too little is ineffective. I make regular copies of my working folder with all the assets (data files, CSS, logos) and the code files themselves, so if I find I need to revert to a previous version, I simply copy that version into my active folder. Eventually, professional solutions might be needed such as GIT, but when starting I believe a notebook is sufficient without needing an additional technical layer to worry about, although I'm sure others would disagree.

CHAPTER 1

Working with Python

In this chapter, we will review some of the essential aspects of software development using Python, such as its data types and powerful data manipulation tools. Python has a way of approaching data management that, while generally intuitive, also has nuances the user must master. For example, if an ordinary person was asked to state the numbers in the range 1 to 5, they would probably answer "1, 2, 3, 4, 5." Unfortunately, in Python, the range(1,5) function returns [1,2,3,4]. Small details like this are obviously important and indicative of how Python approaches data structures. The Python programmer must be comfortable with this "Pythonic Way" through understanding fundamentals like those we provide here for review and reference.

Coding Design: Python and OOD

My first exposure to scientific programming (a long, long time ago) was through developing scientific software for research using FORTRAN. The coding style was very linear – written to process data from the start of a data file to the end, applying various algorithms as appropriate. There was very little interaction between the user and the code – parameters could be set in a configuration file and through command-line arguments, with text messages being output to indicate progress as needed. The advent of GUI interfaces revolutionized programming; displays could provide buttons and sliders and images responding to cursor position.

© Padraig Houlahan 2024
P. Houlahan, *Prototyping Python Dashboards for Scientists and Engineers*,
https://doi.org/10.1007/979-8-8688-0221-8_1

Essentially, this new paradigm demanded a different, nonlinear approach; it required main loops that simply monitored and reacted to events: mouse positions, command keys, and clicks. This was the start of event-driven (reactive) programming.

There was a steep price to be paid for this. Programming became much more complex. It wasn't enough to develop algorithms to number-crunch data, the nonspecialist programmer in a Physics or Biology department also needed to learn how to construct GUI interfaces for graphics display and event handling. A code developed for a PC might break when the OS was upgraded, so different versions might need to be maintained, requiring a relentless search for platform and OS-specific libraries. This was doable with team effort but daunting for the programmer used to linear programming, not wishing to be distracted from their primary pursuits. As languages were coming into fashion such as C and C++, so also was a new philosophy and style of programming, Object-Oriented Design (OOD).

All programs use algorithms to process data. To manage a complex program with hundreds and thousands of lines of code, algorithms are implemented as functions that subdivide the task into manageable pieces. There is a problem though. If a task is subdivided into many subtasks (or functions), how are the results from various functions to be shared as needed with each other? One might think that having all variables be globally accessible to all functions would work. It could – in principle – but with hundreds, perhaps thousands, of variables used in a project, it is too easy for confusion to reign, for an unrelated function working silently to unexpectedly modify values used by another.

A better approach is to pass the variable names as arguments to each function. At least then, in a study of traffic flows, those variables associated with position might be isolated from those associated with motion, and from road conditions if needed. OOD has taken this approach one step further, by defining classes. A class is a definition where a building block of code is constructed with the intention of isolating the functions (methods) designed for the data and the data itself by bundling them together in a single entity. This compartmentalization helps make OOD code very robust and facilitates code sharing. It's a divide and conquer approach, with another tremendous benefit – class variables can be defined that are accessible to all class methods – a kind of local globalization if you will, perhaps not ideal for enormous projects, but incredibly effective for small team research efforts.

OOD naturally allows for managed development; if a coding segment is getting hard to manage and confusing because it is getting too long, then as in linear programming, simply divide it into more methods and classes. Eventually, your objects will be collections of methods and variables. It does seem magical to create an OOD design, define an object in a class, and then what? How is that definition activated? The answer is it is instantiated. If I define class automobile(), then I instantiate it with a command like myCar = automobile(). The color of the car might be myCar. color and the speed found by using a method as in mySpeed = myCar. getCurrentSpeed().

Objects can be used as the basis of other objects, so there is no need to reinvent the wheel. A generic transport class might have methods and variables suitable for Auto, Bike, Train, Ship, and Plane derived classes.

Much has been written about OOD benefits, and I'm barely covering the basics. But I will lastly point out another important aspect – OOD designs are naturally hierarchical and encourage the programmer to look at their project hierarchically. Start with the gross tasks and subdivide and isolate the subtasks and data as needed. While this is also done in linear programming, the problem there is one was often trying to flatten a hierarchical paradigm into a linear one. In my opinion, OOD simply fits better with your mind's innately hierarchical paradigm. This is not to say there is no place for linear programming – for simplicity, it is hard to beat a linear program for building a utility that filters and manipulates data and data files, and I use such utilities extensively – but for more complicated projects, OOD should be the preferred design approach.

Finally, prototyping is an iterative process, where strategies are tried and tested, often rejected, and new approaches attempted based on insights gleaned from previous failures. It makes little sense to try and attempt to replicate this aspect of prototyping, but we can show how we implemented a version of a project's code for you to learn from, to see what coding and design strategies were used and to judge their success by the resulting dashboard. In a subsequent chapter, we will explore the code used to build our prototype, but to properly appreciate what was done, it is necessary to understand coding structures such as Python's data types and how to use powerful solutions such as PLOTLY and DASH. I include many figures with code segments and demonstrations, most of which you can (should?) replicate for yourself using your Python's IDE console in the hope they would help the reader get through some of the more nuanced challenges and have examples to revisit and refer to as necessary.

Python Data Types

[This is probably the section you will most likely be tempted to skip or skim through. It provides an overview of Python's data structures and many code snippets demonstrating how they work. You might be impatient and wish to maintain your momentum while developing code for your project and not wish to be distracted with such arcane details. You can always go online and look up details later. Absolutely. But here's the thing: Python is nuanced and has some surprising quirks – how can such a powerful tool not have? – and having an understanding of its data structures will help you avoid some of those times during code development when you hit a wall; when things don't make sense; when your intuition has failed; and when instead of the intuitive and rational result, you are getting the unexpected. Many times, this will be a consequence of not appreciating how Python data structures function. Remember there is a world of difference between recognizing the familiar ("I've seen this stuff before"), being able to comprehend what is written, and being able to apply the material. Using your Python IDE (coding environment), use the console to mimic the examples here and explore data types created and returned by functions. This section can only provide a short overview, but it will give an important sense of key aspects that must be understood, with many examples of snippets demonstrating key ideas that you can revisit.]

Python has become one of the most popular programming languages used by data scientists and engineers and offers OOD capability and terrific data manipulation solutions, and it has a comprehensive collection of powerful libraries such as Numpy, Scipy, and Pandas available for general use. Many libraries are maintained by experts wishing to contribute to the community at large. Some libraries are commercial, where companies provide their resources and offer tech support to their clients; often, they grant access to noncommercial developers such as faculty and graduate student developers. Of course, if a product emerges with commercial viability, developers need to appropriately license the software used.

A particular feature offered by Python that has become incredibly useful to data managers is a structure called a dataframe. Simply put, a dataframe is a matrix that can hold different kinds of data, numeric or text, and this makes them very useful for holding many kinds of datasets. While simple in concept, there is a fussiness associated with dataframes – we need to be able to slice and dice them, to be able to access blocks and subsets, to be able to index into them, and to use labels and column headers. In this chapter, I will review some of the most powerful techniques for achieving these goals to serve as a convenient reference.

However, before exploring dataframes, it is essential we understand some of the basic data structures such as Python lists, tuples, and sets. In the following sections, I will share an overview of their key features and behaviors.

Lists, Tuples, and Sets

Tuples in Python are simple collections of data that use () while coding. Tuples in Python are immutable with values, once assigned, that cannot be changed. An example is as follows:

```
my_tuple=('apple', 2, 'dog').
```

They are ordered, so my_tuple[2] shows the third element, "dog" in this case.

A set in Python is an unordered collection of unique elements that uses curly braces {} while coding where only one instance of each element is allowed. By unordered, we mean the position of an element is not guaranteed, and we cannot access an element in a set by its index. On the other hand, a list in Python is an ordered collection of values that uses [] brackets while coding, where an element is accessible by its index. (Don't be confused by "ordered" with "sorted!" Ordered means the position of elements is maintained in its sequence.) Figure 1-1 shows how powerful sets can be, where a list with duplicate elements can be turned into a set, and back again, resulting in a list of unique elements. This would be an effective strategy for extracting a unique list of cities from a large database of addresses. Once a set is created, you cannot change the elements, but you can add more (see Figure 1-1).

You cannot access a set element directly, but you can iterate over a set and test if something is a set member.

```
s = {5,4,4,3,3,1,2}                                  # create a set

s
Out[2]: {1, 2, 3, 4, 5}                              # only unique elements used

my_list = ['cat', 'dog', 'mouse', 'dog']            # create a list


my_list
Out[5]: ['cat', 'dog', 'mouse', 'dog']              # all elements used

my_set = set(my_list)                                # create a set from a list

my_set
Out[7]: {'cat', 'dog', 'mouse'}                      # only unique elements kept

my_list2 = list(my_set)                              # create a list from set

my_list2
Out[9]: ['cat', 'dog', 'mouse']                      # now we have a list of unique elements

my_set.add('cow')                                    # add a new item to a set

my_set
Out[11]: {'cat', 'cow', 'dog', 'mouse'}
```

Figure 1-1. *Sets can be constructed from elements listed in {} brackets and will automatically eliminate duplicates. A list with duplicates (my_list) can be converted into a set (my_set) and then converted back, resulting in a list of unique elements (my_list2)*

Lists can be indexed (the first element is [0], the second is [1], and so on) and changed. A list can hold duplicate entries and can have mixed types. Figure 1-2 shows some of the many operations that can be done on lists such as creating, inserting, and removing elements, accessing elements, joining lists, and finding the number of elements present. List elements can be sorted, so if C is a list, do C.sort() or C.reverse() as needed.

```
A = ['cat',True, 5, 8]            # Create list A
A
Out[27]: ['cat', True, 5, 8]

A[2]                              # See 3rd element
Out[28]: 5

len(A)                            # Get the length of A
Out[29]: 4

A[0] = 'dog'                      # Change the first element
A
Out[31]: ['dog', True, 5, 8]

A.append('house')                 # Add an element to A
A
Out[33]: ['dog', True, 5, 8, 'house']

B = ['plane']                     # Create list B

C = A + B                         # Join list A and B
C
Out[36]: ['dog', True, 5, 8, 'house', 'plane']

C.pop(2)                          # Remove the third element
Out[38]: 5
C
Out[39]: ['dog', True, 8, 'house', 'plane']

C.insert(1,'car')                 # Insert an element at position 2
C
Out[41]: ['dog', 'car', True, 8, 'house', 'plane']
```

Figure 1-2. *Elements can be changed, inserted, appended, and removed from lists. Lists can be joined together. Elements can be accessed*

Now here's a tricky one. If you want to copy a list, it is not good enough to do B = A, since B will be a reference to A, and if A changes, so *might* B, depending on whether shallow or deep copying is used by your interpreter. To be on the safe side, **B = A.copy()** is how we save a copy of A to B. After this, we can change or even delete A, but B will remain intact.

The reason why deep and shallow copy options exist is that for a large list that is not changing, the shallow copy is a list of references and is much faster to work with. A deep copy requires more time and space to implement.

Dictionaries

Dictionaries are data collections defined by key/value pairs and use curly bracket notation {}. For example, I might have a dictionary for a student Physics class:

```
my_class = {
            "students": 20,
            "title": "Introduction to Physics",
            "room": 214
}
```

The dictionary keys are found by my_class.keys(). Similarly, I could find the values using my_class.values().

All of these appear very straightforward, but there are pitfalls ahead. At this point, you might guess (wrongly) that if you wanted to access the second dictionary key, it would be a simple list indexing exercise. This is not the case. To illustrate the problem, consider the results shown in Figure 1-3 where I created a dictionary and tried accessing one of the key values. The problem arises because the keys() function returns a Python *set*, not a Python *list*. The solution is to apply the list() function to the **my_class.keys()** output, which returns a list of the dictionary keys, and this list can be indexed. This is shown in Figure 1-4. You can also apply the **list()** function to the output of the values() functions and index the results.

```
my_class = {"room":214, "students": 20, "title": "Physics"}

my_class
Out[3]: {'room': 214, 'students': 20, 'title': 'Physics'}

my_class.keys()
Out[4]: dict_keys(['room', 'students', 'title'])

my_class.values()
Out[5]: dict_values([214, 20, 'Physics'])

k = my_class.keys()

k
Out[7]: dict_keys(['room', 'students', 'title'])

type(k)
Out[8]: dict_keys

k[0]
Traceback (most recent call last):

  Cell In[9], line 1
    k[0]

TypeError: 'dict_keys' object is not subscriptable
```

Figure 1-3. *After creating a dictionary for a class of students, I stored the keys as k = my_class.keys(). I get an error in trying to access the first element using k[0], because k is a set and being unordered cannot be indexed*

```
my_class = {"room":214, "students":20, "title": "Physics"}

my_class
Out[17]: {'room': 214, 'students': 20, 'title': 'Physics'}

k = my_class.keys()

k
Out[19]: dict_keys(['room', 'students', 'title'])

list(k)
Out[20]: ['room', 'students', 'title']

v = my_class.values()

v
Out[22]: dict_values([214, 20, 'Physics'])

list(v)
Out[23]: [214, 20, 'Physics']

list(v)[2]
Out[25]: 'Physics'
```

Figure 1-4. *Applying the list function to the my_class.keys() or my_class.values() output returns a list which can then be indexed*

Series

A Pandas series is like a list, but it allows using labels and integers for indexing, so unlike dictionaries, we use .index() instead of .keys() to get the indexes. We access the Pandas library using a statement like "import pandas as pd," so we can use the shorthand pd.Series to use the Pandas Series() function and so on. Series allows vector operations and can be multiplied by scalers and added together, as demonstrated in Figure 1-5.

```
d = {'a':1, 'cat':5, 4:10}          # Create a dictionary
s = pd.Series(d)                    # Create series s from dictionary d

t = 2*s                             # Multiply all series values by 2

u = t+s                             # Add series s and t together

s                                   # Show s
Out[59]:
a        1
cat      5
4        10
dtype: int64

t                                   # Show t
Out[60]:
a        2
cat      10
4        20
dtype: int64

u                                   # Show u
Out[61]:
a        3
cat      15
4        30
dtype: int64

list(u.values)                      # List the values for series u
Out[66]: [3, 15, 30]

list(u.index)                       # List the indexes for series u
Out[67]: ['a', 'cat', 4]
```

Figure 1-5. *Series can be created from dictionaries, they support vector operations, and they can be added/combined together*

Dataframes

Dataframes are 2D structures that can hold mixed data types. They are created using the Pandas library. Dataframes are very powerful and flexible, and this comes at a cost: there can be many ways to create and manipulate them – some highly nuanced. In what follows, I will demonstrate some of their configuration and manipulation capabilities and some of their not-so-intuitive but important attributes.

Building Dataframes

Dataframes can be created from lists or dictionaries and might have row indexes and/or column names specified.

In Figure 1-6, I show a short code segment creating a dataframe (df) from two lists and the result of running the code in the console. The console shows the code steps and the resulting dataframe.

```
In [100]: import pandas as pd
     ...:
     ...: # make a dataframe using lists
     ...:
     ...: A = [1,2,3]
     ...: B = ['a','b','c']
     ...:
     ...: df = pd.DataFrame(A,B)
     ...: df
Out[100]:
     0
a    1
b    2
c    3
```

Figure 1-6. *Running a code segment that creates a dataframe from two lists*

The result is a little confusing. Yes, the lists have appeared as columns, but there's a zero over the second column and none over the first. Why are they being treated differently? Let's try building the dataframe using a dictionary of the A and B lists instead – again, using our console (see Figure 1-7).

13

```
In [104]:
       ...:
       ...:
       ...: d = {'col1':A, 'col2':B}
       ...:
       ...: df = pd.DataFrame(d)
       ...: df
Out[104]:
   col1 col2
0    1    a
1    2    b
2    3    c
```

Figure 1-7. *Building a dataframe using a dictionary with keywords and a list for each dictionary value. Default indexing appears on the left side*

This result is very interesting. Here, we see the dictionary keyword for each list produces a more intuitive result. We have rows indexed numerically and two columns. In fact, each dictionary key/value pair produces a column in the dataframe. And there is an extra column on the left, showing the index values for each row.

We can change the index column's labels by setting an index parameter to a label list while building the dataframe as in Figure 1-8.

```
       ...:
       ...:
       ...:
       ...: df = pd.DataFrame(d,index=['r1','r2','r3'])
       ...: df
Out[107]:
    col1 col2
r1    1    a
r2    2    b
r3    3    c
```

Figure 1-8. *Creating the dataframe from the dictionary with an index list specified*

14

To understand what happened with the dataframe in Figure 1-6, let's use the dataframe column attribute to explore the columns present in the dataframe produced from lists and that from dictionaries. The results are shown in Figure 1-9 in which the **df.column** output is stored in variables l1 and l2. l2 is a little less cryptic than l1, and to me, this is, frankly, a very annoying result, since not only do we not get a straightforward answer, but we get two different answer styles. The solution is to apply the **len()** function to the variables, and we see the dataframe built from the lists has one column, while that built from the dictionary has two; absent explicit indexing when only using lists, the second list was treated as the index, and so there was only one column of data. This explains the confusion we encountered with Figure 1-6 without the index specified when creating the dataframe, the last list provided was used for the row indexes, and so the resulting dataframe only had one column.

```
In [149]:
     ...: df = pd.DataFrame(A,B)
     ...: l1 = df.columns
     ...:
     ...: df = pd.DataFrame(d)
     ...: l2 = df.columns

In [150]: l1
Out[150]: RangeIndex(start=0, stop=1, step=1)

In [151]: l2
Out[151]: Index(['col1', 'col2'], dtype='object')

In [152]: len(l1)
Out[152]: 1

In [153]: len(l2)
Out[153]: 2
```

Figure 1-9. *Building the dataframe from lists instead of a dictionary with keywords resulted in a one-column dataframe. Regardless of how the dataframe was created here, using len(df.columns) correctly tells us how many columns are present*

We can also explore the row indexes using df.index, but as in the case of df.columns, the output is similarly inconsistent with RangeIndex in one and Index in the other. As a programmer, this is not what I want to see. But there is a good workaround that you can try: regardless of how the dataframe was constructed (list vs. dictionary), apply the list() function to df.columns and df.index. This is shown in Figure 1-10 where the expected output lists are produced.

```
df0 = pd.DataFrame(A,B)

df1 = pd.DataFrame(d)

df0.index
Out[162]: Index(['a', 'b', 'c'], dtype='object')

df1.index
Out[163]: RangeIndex(start=0, stop=3, step=1)

list(df0.index)
Out[164]: ['a', 'b', 'c']

list(df1.index)
Out[165]: [0, 1, 2]

list(df1.columns)
Out[166]: ['col1', 'col2']

list(df0.columns)
Out[167]: [0]
```

Figure 1-10. *Whether the dataframe was produced from lists (df0) or from a dictionary (df1), the list() function shows the row index and column names as a list*

So, at this point, we have row and column labels and the dictionary list data stored in our dataframe columns. The dataframe looks reasonable – there are no confusing columns or missing labels.

Accessing Dataframe Rows and Columns

Our next steps are to see how to access dataframe rows and columns.

Here are two ways regularly found in online articles on how to extract a column from a dataframe using the column names – see Figure 1-11. In the first approach, since the second column name is a simple string ("col2"), I can simply do df.col2. I store the output in variable **e**. I also extract using list notation, df['col2'], and store the result in variable **f**.

However, these results are problematic for a few reasons. First, neither result is a simple list – you might have been expecting – and second, **e** and **f** are of type series – we now need to extract the series data. Yes, this is a little confusing – we extracted series from the dataframe, and we now have to extract lists from the series. We therefore need one last step – use the list() function to convert series data into list form. As shown in Figure 1-11, this yields the desired output: ['a', 'b', 'c'].

```
In [123]:
    ...:
    ...:
    ...: f=df.col2
    ...: e=df['col2']

In [124]: e
Out[124]:
r1    a
r2    b
r3    c
Name: col2, dtype: object

In [125]: f
Out[125]:
r1    a
r2    b
r3    c
Name: col2, dtype: object

In [126]: list(e)
Out[126]: ['a', 'b', 'c']

In [127]: list(f)
Out[127]: ['a', 'b', 'c']

In [128]: type(e)
Out[128]: pandas.core.series.Series

In [130]: type(f)
Out[130]: pandas.core.series.Series
```

Figure 1-11. *Accessing the dataframe's second column using the column name "col2". The results from df.col2 and df['col2'] are both in series form and need to be converted into lists using the list() function*

We should also note that in Figure 1-11, the results of the column extraction operations are of type series – this is why dataframe columns, just like series, can be combined and rescaled vectorially.

Using loc[] and iloc[] to Access by Position

loc[] and iloc[] are two functions that can extract data from dataframes based on position (i.e., using labels and indexes). loc[] uses labels, while iloc[] uses integers. Figure 1-12 shows them being applied to a dataframe to extract the second row ("r2" and index 1). Both functions produce the same overly verbose results, but as before, the len() and list() functions give us the essential information – the actual row contents and size.

```
In [75]: df
Out[75]:
     col1 col2
r1      1    a
r2      2    b
r3      3    c

In [76]: r = df.loc['r2']          # Using .loc[] and label
In [77]: r
Out[77]:
col1    2
col2    b
Name: r2, dtype: object

In [80]: s = df.iloc[1]            # Using .iloc[] with index

In [83]: s
Out[83]:
col1    2
col2    b
Name: r2, dtype: object

len(s)                            # Use len() and list()
Out[84]: 2                        # to get summaries

list(s)
Out[85]: [2, 'b']
```

Figure 1-12. *Using the .loc[] and .iloc[] dataframe attributes on row r2 (index 1) gives the same verbose results from which the actual data list and length can be readily found*

The iloc[] attribute can extract from both rows and columns if its arguments include a comma, that is, it takes the form iloc[row_specifier, column_specifier]. For example, remembering that in Python, a range specified by X:Y starts from X but ends at Y-1, iloc[1:3,0:4] would extract those cells that are found on row indexes 1 and 2 (rows 2 and 3) and columns 0 through 3 as demonstrated in Figure 1-13.

```
df                                    # Our dataframe
Out[2]:
      age   name        pet
r1    22    Art       horse
r2    12    Mary        cat
r3    63    Ken         dog
r2    13    Tim    goldfish

df.iloc[1,1]                          # Extract cell row 2 col 2
Out[3]: 'Mary'

res = df.iloc[0:2,1]                  # Extract cells from rows 1 and 2
res                                   # that are in col 2
Out[7]:
r1      Art
r2     Mary
Name: name, dtype: object

type(res)
Out[8]: pandas.core.series.Series     # A 1 col result is returned as a Series

res = df.iloc[0:2,1:3]                # Extract cells from rows 1 and 2
res                                   # that are also in cols 2 and 3
Out[10]:
     name    pet
r1    Art  horse
r2   Mary    cat

type(res)
Out[11]: pandas.core.frame.DataFrame  # Multi col result is a Dataframe
```

Figure 1-13. *iloc[] can be used to extract subsets of rows and columns using a comma-separated row and column specifiers*

We are not restricted to contiguous rows and columns. Row and column specifiers can be lists. For example, as shown in Figure 1-14, iloc[] can take a list of rows and a list of columns. (If only one list is provided, it defaults to selecting rows.)

```
df                                    # Our dataframe
Out[2]:
      age  name        pet
r1    22   Art       horse
r2    12   Mary        cat
r3    63   Ken         dog
r2    13   Tim    goldfish

df.iloc[[1,3],[0,2]]                  # Row and col specifiers can be lists
Out[12]:
      age        pet
r2    12         cat
r2    13    goldfish
```

Figure 1-14. *Multiple noncontiguous rows and columns can be extracted by using lists of column/row indexes*

Filtering – Extracting Elements by Value

We can create filters to extract information from dataframes based on cell values. For example, **df.age > 20** would identify all rows where the age column was greater than 20. The result is a series of Boolean true or false values. The filter could be saved as a variable, **f = (df.age > 20)**, and passed to the dataframe: **df[f]**. Figure 1-15 demonstrates some examples. Filters can be AND/OR combined using the "&" and "|" operators. To avoid parsing errors, it is best to wrap your filters in parentheses.

Even though the filter is a series, it can be passed to the dataframe, and behind the scenes, Python treats it as a list of true/false elements.

I could have combined my multiple statements into a single one like

df[(df.age > 20) | (df.pet == 'goldfish')].name

which would be very compact. More complicated, compact filters will be harder to read, and it will be a judgment on the programmer's part on how they wish to write their code, whether to strive for compactness or to accept a little more bloat if it clarifies functionality.

21

```
df
Out[37]:
     age  name       pet
r1    22   Art     horse
r2    12  Mary       cat
r3    63   Ken       dog
r2    13   Tim  goldfish


f = (df.age > 20)              # Select by age


f                              # The filter is a boolean series
Out[42]:                       # that identifies rows which satisfy
r1      True                   # the filter's condition
r2     False
r3      True
r2     False
Name: age, dtype: bool


g = (df.pet == 'goldfish')     # Find goldfish owners


df[f | g]                      # Find older than 20 OR goldfish owners
Out[40]:
     age name       pet
r1    22  Art     horse
r3    63  Ken       dog
r2    13  Tim  goldfish


df[f|g].name                   # Get the owner names
Out[41]:
r1     Art
r3     Ken
r2     Tim
Name: name, dtype: object
```

Figure 1-15. *Two filters are applied to the dataframe to find owners either older than 20 or who have a goldfish. Since the result is a dataframe, the df.name construct prints out their names. Note the filters are wrapped in parentheses to avoid parsing errors, and they are stored as variables*

As we can now see, dataframes are part of a data management hierarchy offered by Python and Pandas that provide an environment for a data matrix that includes names and indexes, tools to manipulate the data, and capabilities to extract and filter the data. While it can be surprising to get a series instead of a list when asking for row or column content, and to have the data type silently switch from series to dataframe depending on whether, for example, a single or multiple columns were extracted, Python remains an incredibly useful and flexible tool for working with data, and regularly revisiting summaries and examples like those provided here or online perhaps will expedite your coding.

The Spyder IDE

The Spyder IDE (integrated development environment) presents users with a multi-panel display that is configurable. In Figure 1-16, I show a typical layout where the code is in the left pane, and there is a variable explorer and a console stacked on the right.

Figure 1-16. *The Spyder development environment is a wonderful way to develop code, with many features such as being able to step through code, line by line; being able to explore variable content; and a console for command-line testing*

23

The variable explorer is extremely useful since you can examine the state of a variable or a dataframe and see if they contain the data you think they should! For example, in Figure 1-17, I can see the contents of a dataframe used by the atads.py program developed in a later chapter.

Finally, note that many of the demonstrations exploring Python data types were created using the console window. When you run an application, you can use the console to print various data structures and test out filters before adding them to your code.

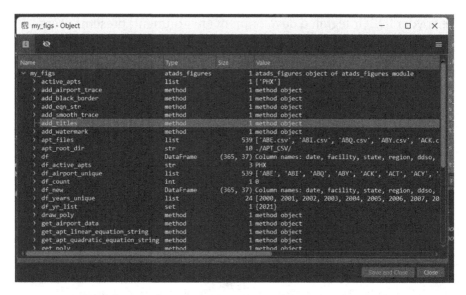

Figure 1-17. *The variable explorer feature allowed me to open an object called **my_figs** and see what some of its dataframes and lists contain. For example, I have 24 entries in my unique year list (**df_years_unique**), and my core dataframe **df** has 365 rows (one for each day of the year) and 37 columns of data. It also shows there are 539 unique airports*

Summary

In this chapter, we reviewed basic Python data structures since seemingly intractable errors often arise when we fail to appreciate how they are defined and used. When coding, using an IDE that allows the user to explore variable and data structures, seeing not just the values but the types and dimensions, makes debugging considerably easier. Being comfortable with lists and dataframes is an essential skill, and this chapter will serve as a quick reference to key techniques and methods.

In the next chapter, we will introduce reactive programming using PLOTLY and DASH. This kind of programming supports interactivity where displayed output responds to widget (e.g., buttons, sliders, checkboxes) selection.

CHAPTER 2

Reactive Programming with PLOTLY and DASH

Reactive programming requires the programmer to build a framework around their core algorithms that supports interactivity, that is, having the application respond to mouse and keyboard input. This framework might constitute the bulk of the programmer's effort; however, the techniques used will often be easily applied to other projects, in effect becoming a mostly reusable (or at least easily customizable) wrapper for the primary function's code. In this chapter, we introduce PLOTLY and DASH to support interactivity and output deployment to a web browser.

Getting Started with PLOTLY

To understand how PLOTLY works, let's explain with a simple dashboard.

© Padraig Houlahan 2024
P. Houlahan, *Prototyping Python Dashboards for Scientists and Engineers*,
https://doi.org/10.1007/979-8-8688-0221-8_2

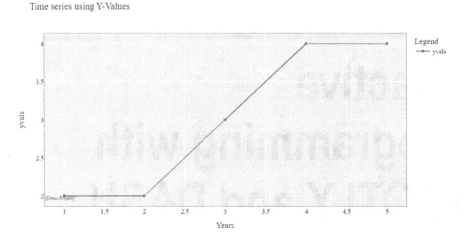

Figure 2-1. *The output created by the code shown in Figure 2-2 appears in a new browser window. Notice the plot is alive and responds to the cursor position.*

A lot is happening in this short code:

1. We import the PLOTLY graphics libraries and the Pandas libraries.

2. The Spyder IDE can be fussy with PLOTLY and be inconsistent in displaying output, but this can be avoided if we use the PLOTLY.io library to set the default browser and use the plot() routine.

3. We build a dataframe from a dictionary based on two lists.

4. We define four utility functions to demonstrate how to add titles, annotation (a watermark), and a border and to apply alternate colors to successive x intervals as a visual cue for the user.

5. We assemble the figure and plot it.

The graphics output appears automatically as a new window that pops up in our browser. PLOTLY embeds buttons at the top-right corner to zoom and move around the plot and for downloading to our desktop. If we resize the window, the plot will also resize. Placing the cursor over a point will show information about the point. PLOTLY does not provide a border around its charts, and because I like to see one, I included a function to do this. Note also I did not have to specify the browser I was using or the operating system. Finally, there might be a bug in the libraries used for this demonstration, so **add_border()** had to be done before the **add_year_block_colors()** routine.

Our example code also introduces us to some of the core PLOTLY graphing routines regularly encountered: **go.Figure()** creates a figure object, while **add_trace()** adds a curve and/or data points; we used **update_layout()** for adding borders and titles and **add_annotation()** for adding text to the chart; finally, we used **add_vrect()** for creating vertical rectangular shapes that could be color coded to create an alternating background used to help delineate years visually. These functions have many more configurations than shown here.

Notice how when working with borders and annotations, we used "paper" coordinates which treat the x and y ranges as going from 0 to 1, regardless of what the actual data values are, so changing the data selected leaves the chart layout unchanged, and also that the plot connects points and adds a marker at each point because we set **mode='lines+markers'** in the **add_trace()** routine.

There are many features built into the demonstration code that you will need to refer to for later work. For now, it is sufficient that the code works and demonstrates some of PLOTLY's many capabilities; it is very likely that a solution already exists for some feature you might wish to implement, and you shouldn't have to reinvent the wheel.

```
import pandas as pd
import plotly.graph_objects as go

import plotly.io as io                                          # Spyder help
io.renderers.default='browser'
from plotly.offline import plot


x = [1,2,3,4,5]
y = [2,2,3,4,4]

d = {'xvals':x, 'yvals':y}
df = pd.DataFrame(d)                                            # build dataframe from dictionary


def add_titles(fig0,title_str,active_variable):                # add title
        fig0.update_layout(
            title=title_str,
            font=dict(family="sans serif",size=14,color="Blue"),
            xaxis_title="Years", yaxis_title=active_variable,
            legend_title="Legend"       )

def add_border(fig0):                                           # add border
        fig0.update_layout(shapes=[go.layout.Shape(
            type='rect',
            xref='paper', yref='paper',
            x0=0., y0=0., x1=1.0, y1=1.0,
            opacity=.4,line={'width': 1, 'color': 'black'})])

def add_watermark(fig0):                                        # add annotation/watermark
        fig0.add_annotation(
            xref="paper",yref="paper",
            x=0, y=0,
            text="[Demo Project]",
            font=dict(family="sans serif",size=10,color="LightSlateGray"),
            showarrow=False, yshift=10)

def add_year_block_colors(fig0,year_min, year_max):            # color code x intervals
    for y in range(int(year_min),int(year_max)):
        if (y % 2) == 0:
            fig0.add_vrect(x0=y, x1=y+1,
                row="all", col=1,
                fillcolor="mistyrose", opacity=0.4, line_width=0)

fig = go.Figure()                                              # build the figure

fig.add_trace(go.Scatter(x=df.xvals, y=df.yvals,
                    mode='lines+markers', name='df yvals',showlegend=True))

add_border(fig)
add_watermark(fig)
add_titles(fig,'Time series for Y-Values','yvals')
add_year_block_colors(fig,x[0],x[4])

plot(fig)
```

Figure 2-2. *A basic PLOTLY program to display a graphic chart. Titles, borders, annotation, and alternate color backgrounds are supported*

But what about OOD? Let's create an OOD version of the code in Figure 2-2 (the code is shown in Figure 2-3). To do so, we will create a class called **my_chart**, which will accept two lists for the x and y variables. For brevity, we will only keep the add_titles() function as a function in the class.

The class will use an initialization function __init__(self, x, y) – note all functions defined in a class must have "self" as their first argument – and it expects two lists to be passed in when invoked. In the initialization, we will set some useful variables, and those we want to reuse are defined with "self." as a prefix. To use class variables, we will invoke them with terminology like self.df.xvals within the class definition, but externally by using the object name, that is, **my_fig.df.xvals**.

```
import pandas as pd
import plotly.graph_objects as go

import plotly.io as io
io.renderers.default='browser'
from plotly.offline import plot


class my_chart:
    def __init__(self, x,y):

        d = {'xvals':x, 'yvals':y}
        self.df = pd.DataFrame(d)                # build dataframe from dictionary
        self.fig = go.Figure()
        self.active_variable='yvals'
        self.title_str='Time series for Y-Values'

        self.fig.add_trace(go.Scatter(x=self.df.xvals, y=self.df.yvals,
                        mode='lines+markers', name='df yvals',showlegend=True))
        self.add_titles()
        plot(self.fig)

    def add_titles(self):
        self.fig.update_layout(
            title=self.title_str,
            font=dict(family="sans serif",size=14,color="Blue"),
            xaxis_title="Years", yaxis_title=self.active_variable,
            legend_title="Legend")


x = [1,2,3,4,5]
y = [2,2,3,4,4]

my_fig = my_chart(x,y)
```

Figure 2-3. *An OOD version of the previous code. Functions and variables are encapsulated in the class definition, and the x and y lists can be passed to the object my_fig instantiated on the last line and immediately create a graph similar to that in Figure 2-1*

31

For this demonstration, for dramatic effect, I build and display the plot while initializing the class, which means that as soon as I create (instantiate) the class object **my_fig**, the plot is immediately displayed. So, in this case, once the class is defined, I only have to change the x-y data lists, and a new chart like that in Figure 2-1 is created after rerunning the program. So, when does a class start working? Answer: As soon as it's instantiated into an object.

Finally, in our OOD design, we could separate the class into an external file that could be reused and shared with others to create a stable and consistent product – one of the major benefits of OOD philosophy.

Getting Started with DASH

DASH is a framework that allows you to create interactive graphics with Python – without having to learn HTML, CSS, or JavaScript. It has its overhead of course; you still have to learn how to implement buttons and sliders and other GUI elements into your dashboard's software, but you are not faced with the chore (stress?) of having to embark on separate JavaScript or CSS projects; it can all be done with Python.

The process of building an interactive dashboard requires us to add capabilities beyond what we achieved in our previous project where we successfully created a web-served graphic. Yes, it was nice that PLOTLY had some useful buttons for scaling and saving the graphics, and even for examining data points with the cursor, but a dashboard needs to allow the user to transform or select or rearrange the data and graphics under the designer's guidance, and so sliders, buttons, checkboxes, and other features (widgets) must be supported. This can all be done using DASH.

To create a DASH application – your dashboard – you need to add instructions defining your dashboard's layout and how to respond to events (callbacks) triggered by actions such as buttons or checkbox clicks. The DASH application will listen for such events, and your code must respond to each type accordingly.

To show how DASH works, let's create a simple application that allows the user to interact with a slider and change a chart output displayed in a browser window. This is a basic dashboard, but if we can do this for one graph and slider, we can add many more; how hard can it be to add a button or a checkbox if we can get a slider to work? Starting with a simple dashboard is an essential first step that puts us on a path to creating more sophisticated ones with many useful chart configurations and input selectors.

For our first dashboard's data, we will create a dataframe using simulated data for a range of years. We will include a layout that recognizes two major components, a chart and a slider. The chart shows the output, while the slider provides the program's input. DASH will automatically detect changes in the slider and handle them with the callback feature and rebuild and refresh the chart in response. The dashboard and code are shown in Figures 2-4 and Figure 2-5, respectively.

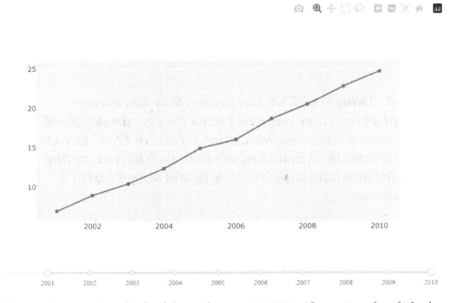

Figure 2-4. *A simple dashboard using DASH. Changing the slider's endpoints changes the displayed chart's year limits*

```
from dash import Dash, dcc, html, Input, Output, callback
import pandas as pd
import random as rd
import plotly.graph_objects as go


list1 = list(range(2001,2011))
list2 = [(2*(x-2000) + 4 + rd.random()) for x in list1]
d = {'year':list1, 'yvals':list2}
df = pd.DataFrame(d)

app = Dash(__name__)

app.layout = html.Div([
    dcc.Graph(id='graph-with-slider'),
    dcc.RangeSlider(2000, 2010, step=1, value=[2001,2010],
        marks={str(year): str(year) for year in df['year'].unique()},
        id='year-slider'
    )
])


@callback(
    Output('graph-with-slider', 'figure'),
    Input('year-slider', 'value'))
def update_figure(selected_years):
    yr1 = selected_years[0]
    yr2 = selected_years[1]

    filtered_df = df[(df.year >= yr1) & (df.year <= yr2)]
    fig = go.Figure()
    fig.add_trace(go.Scatter(x=filtered_df.year, y=filtered_df.yvals,
            mode='lines+markers', name='df yvals',showlegend=False))

    return fig

if __name__ == '__main__':
    app.run(debug=True)
```

Figure 2-5. *Using DASH, we can create a fully functioning dashboard where a user-controlled slider changes the displayed graphic – without requiring JavaScript, HTML, or CSS. The code builds on previously encountered solutions such as constructing and filtering dataframes and creating figures using PLOTLY's graphics objects*

The major parts of the code are as follows:

1. We first import all needed libraries. I also use Python's random library, so I could create a dataset with some randomness – to make it a little more real.

2. A dataframe is constructed from a dictionary using two lists we created.

3. There are entries where a DASH app object is created, an app.layout is defined in which our chart and slider are configured, and a final instruction to run the app.

4. There is a @callback function which handles the input and output to layout components.

5. Immediately after the @callback, there is an update_figure() function in which the figure is built using Plotly graphics object library tools. The resulting figure is returned using a return statement.

We use **dcc** routines from the DASH library to support a **Chart()** and **RangeSlider()**. Both have an essential parameter – an ID – and these are used by the callback's Input/Output specifiers. For the chart, its ID (**graph-with-slider**) is associated as the callback's output – the figure returned by the **update_figure()** function is sent to the chart. The **RangeSlider** has its own ID (**year-slider**) the Input callback uses. Both the callback's Input and Output take two arguments, the ID of the layout component and the DASH data type being transferred. For the chart, the DASH data type is a **"figure"**; for the **RangeSlider**, it's a "value" since a numerical input is used. For the **RangeSlider**, the actual value is passed to the **update_figure()** function and associated with the name set in the update_function()'s argument list.

If there were multiple Input callbacks, they would require a corresponding set of arguments to the **update_function()**, and those arguments must match the Input order.

The callbacks act as a man-in-the-middle directing information received from widgets like sliders into the **update_figure()** function and sending output back to the chart. The IDs are an essential part of making sure information from various widgets and charts are properly routed.

Strictly speaking, the callback( ) is a Python decorator as indicated by the "@" operator. Python decorators are mechanisms that allow you to enhance an existing function, to add something extra, and to decorate it. It's beyond the scope of this book to delve deeply into this very powerful capability, but a simple example is worth exploring – see Figure 2-6. In this example, I create a decorator called **multiply_these**, and its role is to simply add an extra string to the output of the function it is decorating. That function is **a_times_b()** which would normally produce a number. However, when the decorator is applied by placing **@multiply_these** before the **a_times_b()** definition, from that point on, **a_times_b()** will add the extra string before the numeric output. Note that our example successfully passes appropriate arguments through the decorator to the decorated function. Decorators are very powerful and could be used to add a timestamp feature, for example, whenever a particular function is called, without having to edit the function.

```
In [28]:
    ...:
    ...: def multiply_these(func):              # define the decorator
    ...:     def multiply2(x,y):
    ...:         print("Doing multiplication...")   # Add a string to the output
    ...:         return func(x,y)
    ...:
    ...:     return multiply2
    ...:
    ...:
    ...:
    ...: @multiply_these                        #Implement the decorator
    ...: def a_times_b (a,b):
    ...:     print (a*b)

In [29]:
    ...: a_times_b(2,3)                         # Run the function a_times_b()

Doing multiplication...                        # The decorator added the extra string!
6
```

Figure 2-6. *Here, we created a decorator called **multiply_these** and applied it to a simple function (**a_times_b**). Without the decorator, **a_times_b(2,3)** would simply return the number 6, but with the decorator, an extra string "Doing multiplication..." is included in the output*

Another noteworthy detail is that **RangeSlider** returns a list of two values that get assigned to the **selected_years** variable in the **update_ function()** – set to a default of [2001, 2010] – and these values are accessed to define the filter used to select the appropriate years from the dataframe.

We should also note that DASH has its constructs to create HTML, and the code's app.layout creates an HTML <DIV> using the DASH **html.Div** function. So, yes, it is true to say we can build a web-served dashboard without knowing HTML or CSS, but in reality, as we will see in a later chapter, we can make our prototype more attractive (and hence more enticing to the end user) if we use a little CSS to control the page layout.

37

Summary

In this chapter, we covered the basic of how to construct code used to wrap an application's core functionality, so the user can interact with the application. Most importantly, we saw how callbacks track input and layouts and manage the I/O elements.

In the next chapter, we will take the first step in building a dashboard intended to display real-world data used by the aviation industry – namely, solving the problem of how to grab data from an online website and convert it from a format useful to managers (Excel) into one more useful to data scientists (comma-separated values).

CHAPTER 3

Working with Online Data

A chapter on data? Yes, this might seem strange at first. It makes us wonder how hard it can be to work with a TSV (tab-delimited values) or a CSV (comma-separated values) file format. Unfortunately, data files often do not present themselves so nicely. In our case, for our first dashboard, we use the government's ATADS (Air Traffic Activity System) data, and it is accessed by going to a web interface, selecting various configuration options on different pages, and then deciding what format to download – between HTML, Word, and Excel. (Since their audience includes airline and airport managers, this is understandable.) Therefore, it is useful to know how to work with site navigation to grab online data (screen scraping). Even then, there are problems with the ATADS data – we would like to translate it into CSV format and be careful with embedded numeric commas, so "1,234" gets translated into "1234"; extra commas are not good in a CSV file.

About the ATADS Dataset

We will use the Federal Aviation Administration's ATADS dataset. I have used this dataset because I was looking for aviation data to explore and make it available to students, and aviation has always been an interest of mine. As with anyone who likes to work with data, there is always the thrill of seeing if there are hidden features that might be revealed.

© Padraig Houlahan 2024
P. Houlahan, *Prototyping Python Dashboards for Scientists and Engineers*,
https://doi.org/10.1007/979-8-8688-0221-8_3

The ATADS dataset tracks reported flight operations at more than 500 US airports. Airport operators report various activities as being local or itinerant, civil or military, air carrier, air taxi, general aviation, and IFR vs. VFR. The number of each type is reported for each day. Long-term trends can give some indication as to whether an airport's activity is growing or declining and so on, which is why it is of interest to managers and engineers. As previously noted, it can be downloaded in either Excel or HTML format.

To download the data, there are some hoops to jump through. As shown in Figure 3-1, the user is presented with a panel of buttons, each of which sets different configuration parameters, from the download format (HTML, Word, or Excel) to the date ranges to the airport facilities of interest. As a service for managers, the output options can provide rankings and summaries – see Figure 3-2.

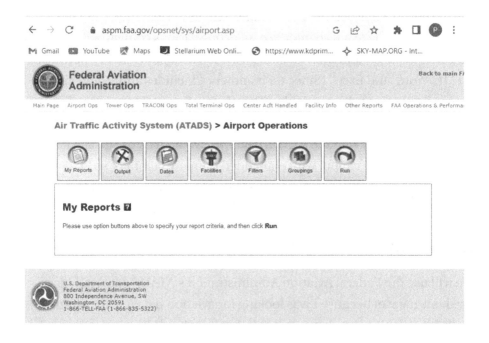

Figure 3-1. *The web page for the ATADS dataset*

We are interested in the raw data, so we will not use summaries and will extract all data for all airports. However, this presents a problem, since the data for the almost 20 years the dataset has been available is quite large – about 170 MB each year. Remember that when encoding data in non-CSV format, there can be lots of extra characters stored in the file. Even the Excel file downloads have HTML tags. To illustrate the issue, consider a highly formatted document with a table of data in HTML format. Individual cells will be wrapped in tags, so the number 42 will appear as <td>42</td> – 11 characters to hold two digits of information! We are interested in the numeric value, and if we can translate this into CSV, we would replace the angle bracket tags with a single comma and use three characters to store the data instead of 11 – a considerable space savings.

But isn't space/storage cheap? Yes and no. Disk storage is indeed relatively inexpensive, but if we wish to manipulate and move gigabytes of data into and out of our program, that's a significant memory usage burden.

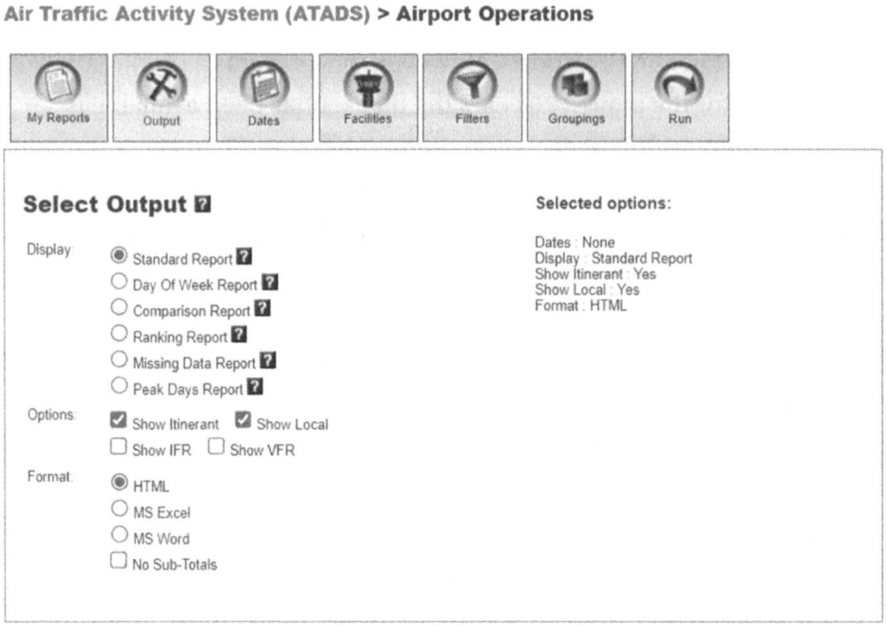

Air Traffic Activity System (ATADS) > Airport Operations

Select Output ▤

Display:
- ● Standard Report ▤
- ○ Day Of Week Report ▤
- ○ Comparison Report ▤
- ○ Ranking Report ▤
- ○ Missing Data Report ▤
- ○ Peak Days Report ▤

Options: ☑ Show Itinerant ☑ Show Local
☐ Show IFR ☐ Show VFR

Format:
- ● HTML
- ○ MS Excel
- ○ MS Word
- ☐ No Sub-Totals

Selected options:

Dates : None
Display : Standard Report
Show Itinerant : Yes
Show Local : Yes
Format : HTML

Figure 3-2. *The output is highly configurable and even offers different comparison and summary report in various data format*

To show what a basic download looks like, I downloaded the data for January 2023 and opened the file using Excel. The result is shown in Figure 3-3. It's a nice presentation, but very far from a CSV style format that can be easily imported into Python.

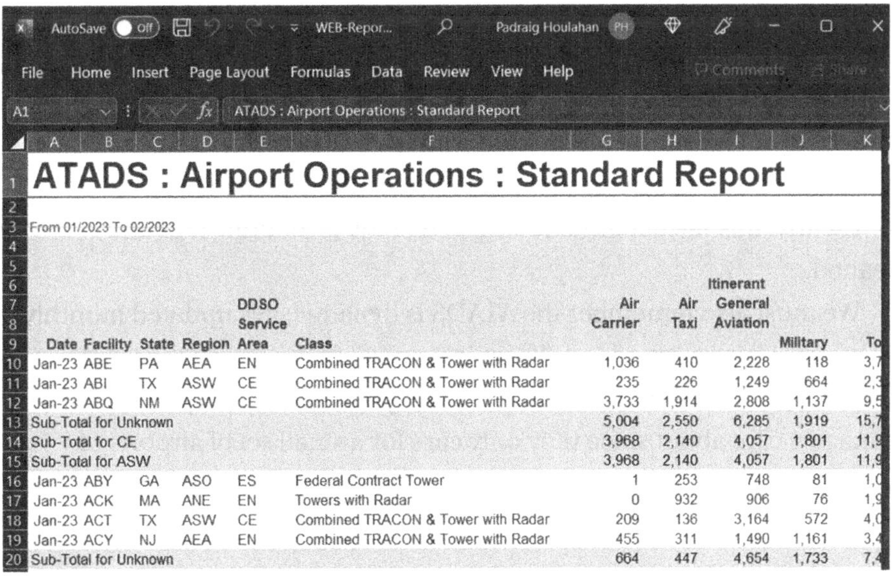

Date	Facility	State	Region	DDSO Service Area	Class	Air Carrier	Air Taxi	Itinerant General Aviation	Military	To
Jan-23	ABE	PA	AEA	EN	Combined TRACON & Tower with Radar	1,036	410	2,228	118	3,7
Jan-23	ABI	TX	ASW	CE	Combined TRACON & Tower with Radar	235	226	1,249	664	2,3
Jan-23	ABQ	NM	ASW	CE	Combined TRACON & Tower with Radar	3,733	1,914	2,808	1,137	9,5
Sub-Total for Unknown						5,004	2,550	6,285	1,919	15,7
Sub-Total for CE						3,968	2,140	4,057	1,801	11,9
Sub-Total for ASW						3,968	2,140	4,057	1,801	11,9
Jan-23	ABY	GA	ASO	ES	Federal Contract Tower	1	253	748	81	1,0
Jan-23	ACK	MA	ANE	EN	Towers with Radar	0	932	906	76	1,9
Jan-23	ACT	TX	ASW	CE	Combined TRACON & Tower with Radar	209	136	3,164	572	4,0
Jan-23	ACY	NJ	AEA	EN	Combined TRACON & Tower with Radar	455	311	1,490	1,161	3,4
Sub-Total for Unknown						664	447	4,654	1,733	7,4

Figure 3-3. *The result of downloading an Excel style formatted report for January 2023 using all airports and viewing in Excel. The spreadsheet is 1644 rows long and contains various headers and labels.*

Let's open the file using Notepad to see what it is really like; see Figures 3-4 and 3-5. The first thing we notice is the file is large - almost 25,000 lines. Figure 3-4 shows the HTML code up to the header on line one of the Excel, while Figure 3-5 shows the raw text corresponding to the Excel spreadsheet entry for Lehigh Valley airport (ABE) Jan-23 (row 10).

We see the HTML encoding is worse than the simple example presented earlier. The numeric count of 1036 is entered as "<td nowrap align=right>1,036</td>" and uses 26 characters to store a four-digit number. And note the number is recorded with a comma, that is, "1,036" and not "1036", which is a problem for CSV files where we expect data to be numeric if we are not careful. (There is a design issue here. Indeed, my CSV file will have a mix of numeric values, such as counts for various categories, and strings for airport labels, but I at least want the count fields/cells to be as clear as possible, and so I want them to be purely numeric, so if I need to open a file with Notebook to visually inspect an entry or to cut a column out of a file with a UNIX command-line utility, I don't want to have to do further data format manipulation.)

Clearly, this file needs to be stripped of the formatting tags and cleaned.

We must also remember the ATADS is dynamic; it is updated monthly when new reports are added. There is no need to download the whole dataset each time. I suspect this is normally not an issue airport and airline managers care about since they only care for a small set of airports of interest.

Figure 3-4. *The plain text version of the downloaded file showing the entries up to the first major label*

```
<tr>
<td nowrap style=text-align:right;>01/2023</td>
<td nowrap style=text-align:left;>ABE </td>
<td nowrap style=text-align:left;>PA</td>
<td nowrap style=text-align:left;>AEA</td>
<td nowrap style=text-align:left;>EN</td>
<td nowrap style=text-align:left;>Combined TRACON & Tower with Radar</td>
<td nowrap align=right>1,036</td>
<td nowrap align=right>410</td>
<td nowrap align=right>2,228</td>
<td nowrap align=right>118</td>
<td style="border-right: #a0a0a0 1px solid;text-align:right;" nowrap>3,792</td>
<td nowrap align=right>2,124</td>
<td nowrap align=right>158</td>
<td style="border-right: #a0a0a0 1px solid;text-align:right;" nowrap>2,282</td>
<td nowrap align=right>6,074</td>
</tr>
```

Figure 3-5. *The segment of the download file Excel used to create row 10 of Figure 3-3. Cell elements are identified because they are wrapped in <td...> ...</td> tags. Note the undesirable numeric commas (1,036 vs. 1036) and the unwanted extra space in line three's "ABE" entry.*

Our data download challenge can now be seen to break down into the following tasks: accessing the data through manual or automated web page navigation (screen scraping); cleaning the data to remove HTML tags, headers, and embedded commas to create a CSV file; and managing the newly created CSV file through adding it to our CSV version of the whole ATADS data used by our application.

ATADS Screen Scraping

It is a chore to have to manually navigate through the different buttons and pages necessary to configure the data download. If this only needs to be done rarely, it's probably easier to simply do it. However, for consistency (i.e., to be sure the same parameter configurations are set each time) and convenience, it would be nice to be able to do this programmatically. In Appendix A, I include a Python code that demonstrates how this might be done. The code uses various libraries and assumes ChromeDriver as a prerequisite. The basic idea is this screen-scraping program will automatically open a Chrome browser, go to the URL with the ATADS page, automatically click the various buttons and links, and finally submit the request. Eventually – depending on how large the data request was – the server will initiate a download to your computer. You will see buttons and checkboxes toggled in the browser window being controlled by the program.

To identify the various parameter names being configured, I used the Chrome browser's Developer Tools. In order to do that, you need to go to the Chrome browser's top-right corner with the three vertical dots, select "More tools," and then select "Developer tools" – please refer to Figure 3-6. For the web page being viewed, the Developer Tools shows its underlying structure – all the HTML and JavaScript tables, divs, variables, and so on.

Figure 3-6. *Finding the Developer Tools in Chrome*

Developer Tools has a window that shows the page elements, and with a little bit of diligence, you can navigate down through the hierarchy, opening collapsed fields, until you find what you want. For example, to find the variable name for the starting month of the date range selection, run the cursor up and down the element list. As you do this, the web page window blocks will highlight. If a block is larger than the element you are interested in, go deeper into the element list (click the small black triangle) and repeat the process. As shown in Figure 3-7, we will eventually find a row in the right panel's hierarchy that exclusively selects the starting month field in the web panel. In there, we find the variable name of interest: "fm_r". This process is repeated for all parameter names of interest. To set a value and click the element, commands like

```
driver.find_element(By.XPATH,"//select[@name='fm_r']
/option[text()='Jan']").click()
```

get the job done. (While I hardwired the selection of January in this example, a more general and elegant solution is warranted.)

With the demo code, all that needs to be done for a later data pull is to change the hardwired data values and rerun the program.

With a little extra effort, this could be run as a scripted crontab under Unix and fully automated if desired, if a virtual desktop was installed on the server. In any event, whether through manual navigation or through the use of screen scraping, we now have a downloaded Excel file that needs to be cleaned and turned into a CSV file.

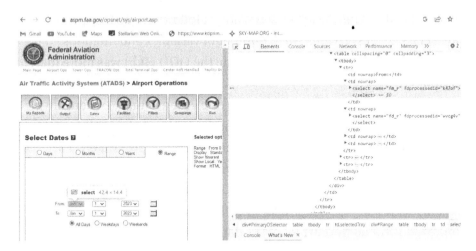

Figure 3-7. *With the cursor highlighted line on the element list on the right, we see the starting month parameter is highlighted on the left, so we now know which element controls the parameter. From the selected element text on the right panel, we see this is controlled by the "fm_r" variable.*

Screen scraping is tricky, but we have a working demo to use and refine and to learn from!

Converting Excel to CSV with Data Cleanup

The downloaded ATADS data file is not in CSV format and has many unwanted elements such as HTML tags that need to be removed. (There is an additional fix that needs to be applied. Airports are assigned three-letter IDs, so Phoenix, for example, is "PHX." However, in the Excel file, it appears as "PHX " – four characters with an extra trailing space that must be removed.) Yes, Excel can save an xls file as a csv, but I chose to write a small utility for this purpose because it can take a long time for Excel to load a large file, and I also needed to apply the customized necessary cleanup fixes.

The code used to do this conversion is shown in the reference section and is called xls2csv.py. It's a straightforward piece of code with a few details I'll address for clarity. First off, note it follows an OOD structure, with a class (xls2csv) with a small number of variables and methods (functions); even in such a small program, the OOD effort was more than helpful.

I'll note that I download ATADS data one year at a time – where I live it can be slow to download larger amounts and can trigger timeouts. I save the downloaded yearly files and rename them appropriately. For example, the downloaded data for 2014 might appear as a file named WEB-Report-12345.xls which I rename to atads2014.xls; these I save in a single directory called ATADS_XLS. If I want to update the data, I simply redownload the current year's data, so the most recent months are included.

The xls2csv.py code reads in an xls file line by line. It skips uninteresting style and header rows, and when it determines it's at the start of the data table, it processes each subsequent row in sequence.

(Remember, an Excel row is a multiline entry in the file – see Figure 3-5.) Cell values are found between "<td>" and "</td>" tags and extracted. Unwanted commas are removed using the Python **re** library pattern substitution capability where any instance of "," is replaced with "".

To fix the extra white space in the airport codes, I used a Python regular expression. Here is the code to remove white spaces using the **re** library in Python:

```
re.sub(r'([A-Z]{3})([ ])',r'\1', line0, count=1)
```

Regular expression (also called regex) is a powerful tool and in this example can be read as follows: when three uppercase characters are encountered [A-Z]{3} followed by a white space [ ], then strip the last character. Only do this once per line (count=1) so as not to corrupt other text fields. This works because the airport three-letter codes appear early in the row. As each cell in a row is processed, an output string is built and eventually written out.

Maintaining the XLS files does present a small challenge since most of the yearly files are static and only the current year is changing. By setting the year_list in xls2csv, you can choose which files to rebuild as necessary.

Managing and Keeping Our Files Up to Date

After processing the XLS files, we now have a collection of CSV files – one for each year in my ATADS_XLS folder – which I store in my ATADS_CSV folder; there are dozens of files since I like to explore long-term trends. The CSV files are more compact than the XLS files, but there's still more than ~300 MB of data. If our ultimate dashboard is to reside on a server to be used by a class of students, having the server loading ~300 MB for each student dashboard is not ideal. A case could be made to import the CSV data into a MySQL server and to have the dashboard grab whatever data it needs, but then we'd need to have such a server running.

We have to make a design decision. To solve our dilemma, I decided to split the data by creating an APT_CSV folder with CSV files for each airport such as JFK.csv, PHX.csv, etc.; since there are about 500 airports, the whole ~300 MB timespan of data is less than 1 MB per airport. Since the dashboard developed in the following chapters assumes users are only exploring a few airports at a time, the application only needs to import few MBs rather than the full dataset – a much more reasonable burden.

The last challenge here is to build a collection of CSV files – one for each airport – from the collection of annual CSV files, and this brings us to another design choice. Do we want to be efficient and simply append the latest data to the existing airport CSV files or do a clean rebuild from scratch? The problem here is that when we download the annual ATADS data, say, for 2020 June, the APT_CSV collection already has 2020's April data. I decided it would be less confusing, albeit more inefficient, to simply rebuild the APT_CSV folder contents from the ATADS_CSV annual files, to sacrifice a little efficiency to maintain clarity and robustness.

Split_by_apt.py is a simple program that achieves these goals. For each year's file in the ATADS_CSV folder, it creates a list of unique airport IDs. For each airport, it appends that airport's entries to the corresponding airport file in the APT_CSV folder. It uses the powerful Pandas library. The code reads a year's CSV file into a master dataframe and uses the master dataframe's second column to create a unique list of airports. For each airport, it creates a secondary dataframe for that airport and appends it to the airport's csv files in APT_CSV.

The APT_CSV folder is the final result of our data importation, cleanup, and conversion from Excel to CSV. It contains a CSV file for every airport in the ATADS dataset, and these files can now be used as input for our dashboards.

Summary

In practice, the programmer might not receive data in a convenient format or have it be easily accessible. Since many datasets are now shared over the Internet, in this chapter we showed a solution that allowed us to download ATADS data – a solution that could be added to a unix chron file, so the process could be automated. For our data, we also encountered a problem in that it was not in CSV format, but rather in Excel. Because our files are so large, Excel could be very slow if we used it to do the conversion. In any case, we developed a Python code to do the format conversion into CSV, and this code could also be run as part of a unix chron job.

In the next chapter, we address some of the design issues our ATADS dashboard will raise, such as which features we would like to include and how the application might be deployed so it can be accessed by remote users.

CHAPTER 4

Planning the Dashboard Prototype

In this chapter, we go deeper into technical design issues and list the main tasks our project faces. We also take the opportunity to review linear equations, since we will use these for trend analysis and forecasting.

Overview

For this project, we want to use a government dataset (the ATADS) that tracks daily airport operations counts (including many metrics such as how many flights were commercial or military or conducted under visual or instrument conditions) at about 500 US airports, and we wish to make the data accessible to end users. By accessible, I mean we would like the end users (airport and airline managers, aviation and business researchers, and specialists) to be able to quickly see what's happening with various aspects of selected airport operations. We would like the end user to be able to select among airports and time ranges, to see some detailed metrics, and to be able to produce nice charts for reports and web content. Interactivity would be an essential component to support user explorations. Such access would allow users to compare airports, to identify trends, and to illuminate planning.

© Padraig Houlahan 2024

P. Houlahan, *Prototyping Python Dashboards for Scientists and Engineers*,
https://doi.org/10.1007/979-8-8688-0221-8_4

Interactivity entails programming that allows users to click menus and buttons and to select graphical regions for deeper study. This is a dynamic process and requires a reactive programming environment, such as the one that PLOTLY/DASH provides.

But we also need to solve the distribution problem – making our solution available to others, perhaps at remote parts of the Internet. This we will do by adapting our software into a services kind of environment that our Nginx web server will use to respond to remote requests.

Our goals therefore break down into the following elements: data import and manipulation, reactive code development, creation of Linux services, and the creation of a Unix web server such as Nginx. In addition, to support end users we will create a web portal that uses blogging software, so we have a centralized environment for feedback, support, and documentation – essential resources for our users.

Figure 4-1 shows the primary relationships between the project's elements.

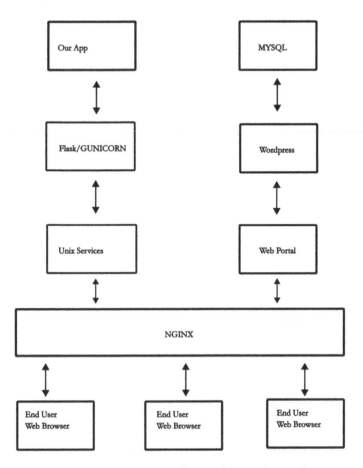

Figure 4-1. *Our project overview shows the various elements we will bring together, so we can share our solution and provide a central resource and documentation area*

Here is what Figure 4-1 is telling us: we will develop a program (for now, called "Our App") that needs to be shared. Our Unix operating system will host critical servers we will need to configure. We will use the Nginx web server to respond to incoming requests from remote user web browsers. Our web portal is simply an HTML/CSS page with links to the WordPress server. For logins and document tracking, WordPress uses a MySQL database server that we also need to configure. Flask and GUNICORN are two solutions that allow our systems to transact queries and responses from our application to meet end-user needs.

Since we are working with aviation data, an analogy will help us understand the Flask/GUNICORN role. In very rough terms, let's consider our project to be an airport, and our goal is to move cargo and passengers in and out of this airport. This movement of goods and people is like servicing the requests and responses that our project's application handles. Flask is like a small single-engine aircraft that can carry a small number of passengers. It can get the job done, but it's fragile and not suitable for heavy weather conditions or large cargo surges. GUNICORN, on the other hand, is like a large commercial jet; it can handle many passengers and lots of cargo under less-than-ideal conditions. During development, we will use Flask to get our project working, but we will switch to GUNICORN when we are ready to put our project into production.

You might be wondering which of the blocks in Figure 4-1 is our actual server. The answer is, other than the row of remote user browsers, all of the blocks are servers we require in some sense, and that group as a whole is the system that services our ATADS users. In my case, I have them all on a single virtual platform hosting an Ubuntu operating system, but if the project needed to grow (perhaps to delegate control to colleagues), I could move some blocks to other virtual systems. Also, I will note there is an advantage in having all the blocks on one virtual server since that server could be replicated as many times as necessary to offer scalability and redundancy.

Now it's time to take a slow deep breath. Yes, our project is complicated, but there is good news. Other than for our app, we are not breaking new ground here; most blocks/components have all been done before. And because of either benign enlightenment or sheer exasperation with customers tormenting their customer support teams, many of the server farm vendors that can host such services relatively inexpensively provide remarkably detailed step-by-step instructions on how to do each piece. We, therefore, can direct our creativity into building the core application and can use well-documented solutions for implementing all the Flask/GUNICORN, MySQL, WordPress, and Nginx installation and configuration. Yes, it will be fussy, but we will have great templates to guide us.

Project Tasks

Our first dashboard should offer the user an interface with which they can readily explore the underlying data. This means we have to carefully find a balance between offering them too much information that confuses and clutters the presentation and having a display that's too simplistic.

Since our data is essentially a collection of time series, it is natural that we would want to see data plotted against a time axis, with an ability to concentrate on different year intervals. There are multiple kinds of flight operations, and being able to select between them would also be desirable. Being able to select multiple airports would be useful for studies interested in identifying and seeing how peer airports compare.

Being able to display raw data is not enough; we would also like to be able to declutter the data and see underlying trends which will require polynomial regressions and smoothing.

Our prototype will run on a personal computer, and once this is working, we can address issues concerning how to host it in a UNIX environment for sharing. We can always add more features and panels and charts later, but, first, we need to display our ATADS data and build the tools needed to modify our chart based on user-selected options using PLOTLY, and this means we must now specify what exactly we would like our dashboard to do.

No matter what data is being used, it cannot be overstated how important it is for the user to have a sense of what it looks like, and being able to see it in graphical form is extremely important. Aviation researchers and professionals need to compare airports – to identify peer airports, to see how similar ones are performing, to quickly identify features and patterns, and to quantify trends for forecasting discussions.

So, here's a list of dashboard capabilities we will offer the end user:

1. An ability to select one or more airports for study.

2. A means to select a range of years, so either long-term or short-term trends can be studied.

3. An ability to select one of the operational metrics (ATADS data columns).

4. A graphical display of the data's time series, so patterns and trends can be identified.

5. An option to smooth data so longer-term features are more easily seen and to reduce short-period clutter.

6. If desired, graphs for different airports should overlap each other to make it easier to compare them.

7. A means to display basic trends (linear and quadratic) and their associated equations.

8. An ability to explore and document periodicities in the data since often very well-defined patterns, weekly and seasonal, are present and it would be useful to quantify them.

It should go without saying that we would like to achieve these goals with a dashboard that is convenient and pleasing to use, so it should be intuitive and attractive. You might not get the correct balance between elegance and usefulness, between simplicity and clutter, the first time, but that's what prototyping is all about. Our goal is to get the first prototype version working which will serve as a baseline for later adjustments.

Trends and Forecasts

Having a database is pointless if we don't know how to use it effectively. While it is true that data is sometimes (often?) collected with no real underlying sense of what it might be used for, other than perhaps justifying the collection effort and demonstrating that something is being *done*, in general, data is collected to help illuminate a subject. Professionals will use data to help them better understand what is going on and to appreciate patterns and trends, and so data can help them anticipate the future. One of the major goals of our dashboard is to help with the visualization process; the other is to help identify patterns and trends and facilitate forecasting.

For our purposes, since we are mainly concerned with time series data, we will simply say that a trend is a curve we distill from the data. For us, we will only care about low-order polynomial curves (equations involving terms like t or t^2) for a variety of reasons. First, polynomials are a straightforward type of math function that is user-friendly and can be applied to various sets of data. Unlike more complex functions like those found in trigonometry, polynomials are easy for regular users to understand and work with. Now, when it comes to the order of polynomials, higher-order ones can get confusing, especially for those who are not dedicated researchers. For our data, the most useful equations are the simple first-order, linear ones, which represent straight lines. Whether it's a linear or higher-order equation, having a mathematical equation is valuable because it not only helps us measure current behaviors but also enables us to predict future ones. Linear (first order) and quadratic (second order) polynomials are equations of the form

$$y = a_0 + a_1 \times t,$$

Eqn 4.1

and

$$y = a_0 + a_1 \times t + a_2 \times t^2$$

Eqn 4.2

respectively. The highest power is the order of the polynomial. In equations like these, the a_0, a_1, a_2 are numbers called the coefficients. Curve fitting (sometimes called "regression") is a set of mathematical techniques applied to a dataset used to determine the coefficients that characterize the most reasonable curve that passes through the data. The distances between the data points and the curve are kept to a minimum. This can involve complicated mathematics but has been encapsulated in the Python **poly** libraries we will use.

Note, when comparing the coefficients derived for a linear curve fit with those for the quadratic, there is absolutely no requirement the a_0 and a_1 coefficients will match or that you can simply add another term to an existing linear solution and get the quadratic. Once the coefficients are found (linear or quadratic), their associated curve is generated using either Equation 4.1 or Equation 4.2 as appropriate.

In Equation 5.1, the a_1 coefficient is the slope (or trend or rate) of the line.

In Equation 5.2, the slope (from calculus) would be $a_1 + 2 \times a_2 \times t$ – and changes with t!

In both equations, a_0 is the intercept, the value when t is 0.

A benefit of curve fitting is that, mathematically, it can distill the essence of a large collection of points down to two or three numbers, the coefficients. It is beyond the scope of this book to delve deeper into this topic, but be extremely careful here. At the same time, you can always apply curve fitting algorithms and perhaps even get better fits using higher-order polynomials; their interpretation can be fraught with danger. This was one of the main reasons I avoided adding higher polynomial orders as a dashboard option. Having said this, being intimately familiar with your data's behavior and trends will make curve fitting interpretation more reasonable. This kind of familiarity is gained, in part, by exploring your data visually with tools like our dashboard!

Other Design Considerations

Now that we have an initial set of goals/tasks to aim for, we can begin to visualize how we might achieve them. Keeping it simple (to begin with), we at least want a chart, so trends can be studied, and we need a means to select various options, so there are graphical tasks and layout ones. We have seen how we could benefit from using an OOD approach which would help keep the project organized, and we will take it one step further by splitting the code across multiple files. (Yes, a dedicated programmer might argue our project is small enough we could leave it as a single file – the first draft is only about 800 lines of code – but projects grow, and later versions are longer, and there is a point where it's tiresome, visually overwhelming, and distracting when having to scroll or navigate around hundreds of lines of code.) Referring back to our example of a DASH application in Figure 2-5, a natural division of work suggests itself – there is a section where layouts are handled, there is a section inside the update function where figures are created, and of course there is the whole DASH solution itself. So, our prototype will use three files, one for the top level (similar to Figure 2-5) that oversees everything and one each supporting the layout and figure creation: atads.py, atads_figures.py, and atads_layout.py.

atads_figures.py has a class (atads) that handles the data import and figure creation, while atads_layout.py has a single class (my_layout) that supports widgets like drop-down menus, checkboxes, and radio buttons and allows the end user to select airports and to configure other parameters. So, not only are we isolating code into separate files for convenience and to reduce clutter, but we are also encapsulating using Python classes, so project tasks are self-contained and more stable.

Summary

In this chapter, we described our project in greater detail, establishing a list of desirable features and capabilities, with a discussion on trends and polynomial equations, which empower users by providing concrete summaries of airport performance and which also facilitate comparing airports with one another.

In the next chapter, we will start building a simple dashboard, which will serve as the first prototype with a single chart and some widgets that will allow us to see how PLOTLY and DASH work. This will be where we explicitly build an OOD solution following our design goals. Once the first prototype is working, it is easier to add additional charts and features.

CHAPTER 5

Our First Dashboard

In this chapter, we develop a dashboard with working widgets, organized into three files to implement an OOD solution that will support future refinement and maintainability. We will use class constructs to group figure rendering and data management, layout, and reactive programming together. Core algorithms will be used to support regression/trend analysis and CSS code shown to allow the programmer better control chart and widget screen layout. The result will be a basic but fully functional dashboard.

Figure 5-1 shows us our initial dashboard showing data for three years near the pandemic. The dashboard consists of two main panels – a configuration side to select among airports, years, and display options (including whether to apply linear or quadratic curve fitting) and a panel showing the plotted data and other summaries. Unlike the example in Figure 2-5, where the slider appeared below the chart, this layout uses a CSS file which we will address later to arrange the panels side by side. We have attempted to produce a dashboard that has highly configurable, useful (intuitive), and pleasing graphics. Because others might like to use the dashboard, a watermark is included, so an airport publishing web content could help ensure subsequent inquiries are directed back to them.

Pandemic effects are apparent, and we used smoothing (red curve). A linear curve was fitted that tells us that over this interval, these daily operations increased by 122.3 annually. A gentle alternating color background helps delineate yearly intervals, and the mouseover shows the actual, smoothed, and curve-fitted counts for the day.

© Padraig Houlahan 2024
P. Houlahan, *Prototyping Python Dashboards for Scientists and Engineers*,
https://doi.org/10.1007/979-8-8688-0221-8_5

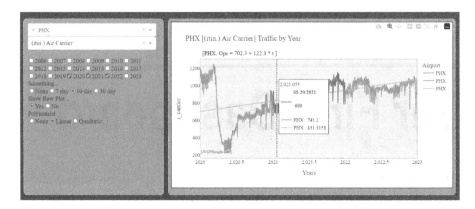

Figure 5-1. *Our first dashboard shows daily Air Carrier traffic operations at Phoenix Sky Harbor airport. The user can select among a variety of options on the left to select the data used to create the chart.*

We are essentially following the design we discussed in the introduction of Chapter 2 to DASH programming and design considerations of Chapter 4, and we have already demonstrated all of the techniques needed to produce the dashboard, so we will now go through this dashboard's specifics by exploring its three core files – atads.py, atads_figures.py, and atads_layout.py – and the CSS file used to finesse the on-screen layout.

The atads.py File

Our **atads.py** file (shown in Figure 5-2) is our prototype's top-level file – similar to the DASH example shown in Figure 2-6. Our prototype follows the same overall flow: a DASH app is created using app = dash.Dash(), then the layout defined with **app.layout()**, the callbacks set, and the app activated (**app.run**).

We created two new classes. The first **atads_layout()** handles the widgets used by the user to specify airports, years, and graphing options, while the other **atads_figures()** handles the graphing and charting tasks. This not only allows us to group our tasks functionally but also reduces clutter since the new classes can be kept in external files.

Since there are many configuration options, individual settings were grouped into the **atads_layout.controls()** method; this is where the drop-down menu, year selection, and plotting parameters that appear in the dashboard's settings panel are configured. DASH constructs, such as **html.Div**, **html.Label**, and **html.Br**, are used to control the control panel layout and to provide labels for various parameter groups. Some of the controls need data stored in the **my_figs** object to create the pull-down menus, and these are passed as arguments to the **my_layout.controls()** method; even though the **atads_layout** class uses information from the **atads_figures** class, so little information is passed between them, it was simpler to do it this way instead of using Python inheritance, that is, basing one class on another.

The callback section is more complex because while we only have one output (the main chart), we have more user-selectable settings, and each needs its own callback **Input()** statement. The order of the **Input()** lines must match the order of the variables presented to the **update_dashboard()** function. In DASH, callback **Input()** and **Output()** statements map to the update's command-line arguments and **return** value, respectively. Finally, **update_dashboard()** updates the data and the displayed chart if changes are made to the settings.

While our design has some limitations (e.g., it would be better to split the callbacks, so airport data was not refreshed whenever parameters were changed, and, perhaps, we should hide the **change_settings()** inside the layout class), our design is simple and easily modified. If we wanted to add additional parameter options, we can see we would only need to add an appropriate **Input()** for the callbacks, an appropriate argument to the **update_dashboard()**, and an entry to the **atads_layout.controls()** function.

```
import dash
from dash.dependencies import Input, Output
from atads_layout import  atads_layout
from atads_figures import atads_figures
from dash import html

app = dash.Dash(__name__)
my_figs   = atads_figures()
my_layout = atads_layout()


app.layout = html.Div(
            className="content",
            children=[
            #configure_settings(),
            my_layout.controls(
                my_figs.df_airport_unique,
                my_figs.var_dict,
                my_figs.df_years_unique),
            my_layout.chart()
            ])

@app.callback(
    Output("scatter_plot", "figure"),
    Input("airports",      "value"),
    Input("years",         "value"),
    Input("use_variable",  "value"),
    Input("smoothing",     "value"),
    Input("show_raw",      "value"),
    Input("show_poly",     "value")
    )
def update_dashboard(airport_list, yr_list, active_variable, smoothing,show_raw,show_poly):
    my_figs.get_airport_data(airport_list,yr_list)
    my_figs.update_mainchart(airport_list, yr_list, active_variable, smoothing,show_raw,show_poly)
    return  my_figs.fig_main

if __name__ == "__main__":
    app.run(host='0.0.0.0')
```

Figure 5-2. *The **atads.py** file uses two new classes (**atads_layout** and **atads_figures**) to manage layout and charting. This division of labor allows this top-level code to be simpler since the extra classes are imported from external files. A key method for managing user inputs – **my_layout.controls()** – accepts critical data from the **my_figs** object through its function argument list.*

The atads_layout Class

Our **atads_layout.py** file contains the **atads_layout** class definition. This class is needed to associate various widgets and the output chart to the callbacks, where each association is done by one of the class methods. The class methods are functions that return a DASH function; all the work is done by a single line of code in each method. To ensure the association is mapped correctly, each function sets an ID for its widget – used by a top-level callback.

Figures 5-3a and 5-3b show the **atads_layout.py** file's contents. In Figure 5-3a, widgets for displaying the main chart and user-selectable options are shown. The **dropdown_airports()** method expects to receive a list of unique airports, while the **dropdown_use_variable()** expects a dictionary, where each member maps a user-friendly variable name to an internal variable.

```
from dash import dcc
from dash import html

class atads_layout:
    def controls(self,df_airport_unique,var_dict,df_years_unique):
        return html.Div(
            className="parameter_selections",
            children=
            [
              self.dropdown_airports(df_airport_unique),
              self.dropdown_use_variable(var_dict),
              html.Br(),
              self.select_years(df_years_unique),
              html.Label('Smoothing...'),
              self.radio_smoothing(),
              html.Label('Show Raw Plot...'),
              self.radio_show_raw(),
              html.Label('Polynomial'),
              self.radio_show_poly()
          ])

    def chart(self):
        return  html.Div(
                className="chart",
                children=[
                dcc.Graph(id="scatter_plot"),
            ])

    def dropdown_airports(self,df_airport_unique):
        return dcc.Dropdown(id='airports', options=[
                {'label': i, 'value': i} for i in df_airport_unique
                ], multi=True, value=["PHX"],placeholder='Filter by airport...')

    def dropdown_use_variable(self,var_dict):
        return    dcc.Dropdown(id='use_variable',
                    options=[
                    {'label':var_dict[i], 'value':i} for i in var_dict

                ], multi=False, value="i_carrier",placeholder='Filter by variable...')

    def select_years(self,df_years_unique):
        return dcc.Checklist(id='years', options=[
                {'label': i, 'value': i} for i in df_years_unique

                ],inline=True, value=[2023])
```

Figure 5-3a. *These methods from the **atads_layout()** class manage the user selection display and inputs for the drop-down menus and year-selection checkboxes.*

In Figure 5-3b, the remaining methods control the radio button widgets used to specify whether the raw chart should be displayed and/or a smoothed version and/or a polynomial (linear or quadratic) fitted curve.

Our **atads_layout** class demonstrates radio buttons, checkboxes, and pull-down menus and is intuitively extendible, and its methods serve as templates for adding more selectable parameters as the prototype dashboard evolves.

```python
def radio_smoothing(self):
    return   dcc.RadioItems(id='smoothing',
                    options=[
                        {'label': 'None', 'value': '0'},
                        {'label': '7 day', 'value': '7'},
                        {'label': '10 day', 'value': '10'},
                        {'label': '30 day', 'value': '30'}
                    ],
                    value='0',
                    labelStyle={'display': 'inline-block'}
                )


def radio_show_raw(self):
        return  dcc.RadioItems(id='show_raw',
                options=[
                    {'label': 'Yes', 'value': '1'},
                    {'label': 'No', 'value': '0'},

                ],
                value='1',
                labelStyle={'display': 'inline-block'}
            )

def radio_show_poly(self):
    return      dcc.RadioItems(id='show_poly',
                options=[
                    {'label': 'None', 'value': '0'},
                    {'label': 'Linear', 'value': '1'},
                    {'label': 'Quadratic', 'value': '2'}
                ],
                value='0',
                labelStyle={'display': 'inline-block'}
            )
```

Figure 5-3b. *The radio buttons used by the prototype are also defined in the **atads_layout** class.*

The atads_figures Class

The **atads_figures** class is the heart of our dashboard's development effort; this is where we manipulate our data and build original algorithms tailored to our goals. Conditional on user selections, our **atads_figures** class handles how data is imported and adjusted, filtered, processed, and the desired graphical outputs created.

To understand how the **atads_figures** class functions, we will first review how its initialization works including managing data and variable names, then the **update_maincart()** method which controls how the main scatter plot is rendered, and then the remaining methods grouped by purpose, enhancing the chart's appearance, adding line traces, and creating polynomial curve fits.

Initialization

Importing and adjusting the data consists of inputting data from our APT_ CSV folder (see Chapter 3), where the CSV files for our airports of interest are read using the Pandas read_csv() function and saved as a dataframe **self.df[]**. But since the CSV data uses separate columns for years, months, and days, we also create a new column **ydecimal** where each row's year, day, and month, converted into a decimal year format, are stored to facilitate plotting time series.

The class initialization breaks down into two main parts: variable name management and initialization of miscellaneous variables.

Variable Name Management

Before we can work with our data, we must define what names to assign our CSV data columns, and we need a dictionary to display more friendly variable names to the user instead of internally succinct ones, so, for example, in the variable selection drop-down menu, the user might select **IFR (itin.) Air Carrier** and that would be mapped to **ifri_carrier**.

We use **self.names[]** to set the initial column names and then **self. var_names**[] and **self.var_labels**[] for building the variable selection's menu name mapping dictionary (**self.var_dict**[]). The dictionary is created by **self.make_var_dict**(). (There is some redundancy here – born out of a desire to keep options open when working with other datasets.) Note that not all data columns were of interest, and some were excluded when making the dictionary. (See Figure 5-4.)

```
import pandas as pd
import plotly.graph_objects as go
import numpy.polynomial.polynomial as poly
import numpy as np
from os import listdir
from os.path import isfile, join

class atads_figures:

    def __init__(self):

        self.names=[
                        "date", "facility", "state", "region", "ddso", "class",
                        "ifri_carrier", "ifri_taxi","ifri_general","ifri_mil","ifri_total",
                        "vfri_carrier", "vfri_taxi","vfri_general","vfri_mil","vfri_total",
                        "i_carrier", "i_taxi","i_general","i_mil","i_total",
                        "loc_civ", "loc_mil", "loc_total",
                        "total_ops",
                        "j1","j2","j3","j4","j5","j6"
                        ]
        self.var_names=[

                        "ifri_carrier", "ifri_taxi","ifri_general","ifri_mil","ifri_total",
                        "vfri_carrier", "vfri_taxi","vfri_general","vfri_mil","vfri_total",
                        "i_carrier", "i_taxi","i_general","i_mil","i_total",
                        "loc_civ", "loc_mil", "loc_total",
                        "total_ops"
                        ]

        self.var_labels=[
                        "IFR (itin.) Air Carrier",
                        "IFR (itin.) Air Taxi",
                        "IFR (itin.) Gen. Av.",
                        "IFR (itin.) Military",
                        "IFR (itin.) Total",
                        "VFR (itin.) Air Carrier",
                        "VFR (itin.) Air Taxi",
                        "VFR (itin.) Gen. Av.",
                        "VFR (itin.) Military",
                        "VFR (itin.) Total",
                        "(itin.) Air Carrier",
                        "(itin.) Air Taxi",
                        "(itin.) Gen. Av.",
                        "(itin.) Military",
                        "(itin.) Total",
                        "(local) Civilian",
                        "(local) Military",
                        "(local) Total",
                        "Total Ops."
                        ]
```

Figure 5-4. The **atads_figures** class initialization customizes the variable names used for the CSV data (self.names) and also defines the lists used for creating the **self.var_dict** used to map user-friendly (**self.var_labels**) and the more succinct forms (**self.var_names**) used internally.

Miscellaneous Variable Initialization

At the end of the __init__() function (see Figure 5-5), some useful variables are set, the variable name dictionary created, a list of unique airports built, and data for PHX (Phoenix Sky Harbor) read as a default.

```
self.active_apts = ['PHX']
self.apt_root_dir='./APT_CSV/'
self.my_debug='no bug report'
self.df = pd.DataFrame()
self.df_count = 0
self.df_years_unique = [i for i in range(2006,2024, 1)]

self.make_var_dict()
self.get_airport_list()
self.get_airport_data(['PHX'],{2021})
```

Figure 5-5. *The last part of the **atads_figures** class initialization sets some general variables and builds the **self.var_dict** dictionary, builds a list of available airports, and reads in PHX as the default airport.*

Class Methods

The class methods, other than __init__(), naturally form into different groups based on their roles:

1. A set of utilities supporting data input and managing the main internal data repository – the **self.df** dataframe

2. The **update_mainchart()** used to create the graphical output

3. Utilities to draw line graphics

4. Utilities to support linear and quadratic polynomial curve fitting

5. Utilities to enhance graphical output – titles, watermarks, colors, etc.

I/O and Variable Name Utilities

(See Figure 5-6.)

make_var_dict() is used for building the variable name dictionary (**self.var_dict**) needed by the variable selection drop-down menu, so user-friendly menu items are mapped onto the more terse variables used by the code. The method uses the very useful **zip()** utility for creating a dictionary from two lists.

get_airport_list() reads the contents in **self.apt_root_dir** ("/APT_CSV"). Since there is a CSV file for each airport, the contents are used to create a list of unique airports. (Note the use of the **listdir**, **isfile**, and **join** functions imported from the Python **os** libraries.)

get_airport_data() is used for importing data for our selected airport list. At the start, it empties the **self.df** dataframe using **iloc[0:0]** and appends the output (**self.df_new**) from each **read_apt()** call to **self. df**. **read_apt()** is run for each airport on the selected airport list. After the method runs, all required airport data is available in the **self.df[]** dataframe for use throughout the rest of the application.

The **read_apt()** method reads one airport at a time into a dataframe. In addition, a column is added to the dataframe for a new variable (**ydecimal**) which is the date expressed in decimal years. The method uses the Pandas csv library function **read_csv()** to load csv data into a dataframe. The dataframe is filtered using the **isin()** function to extract only entries for years listed.

```
def make_var_dict(self):
    keys_list = self.var_names
    values_list = self.var_labels
    zip_iterator = zip(keys_list, values_list)
    self.var_dict = dict(zip_iterator)

def get_airport_list(self):
    self.df_airport_unique=[]
    self.apt_files = [f for f in listdir(self.apt_root_dir) if isfile(join(self.apt_root_dir, f))]
    for f in self.apt_files:
        first_chars = f[0:3]
        self.df_airport_unique.append(first_chars)
    self.df_airport_unique.sort()

def get_airport_data(self,apt_list, yr_list):
    self.df = self.df.iloc[0:0]
    for i in apt_list:
        self.read_apt(i, yr_list)
        self.df = pd.concat([self.df,self.df_new])

def read_apt(self,apt, yr_list):
    filename=self.apt_root_dir+apt+'.csv'              |
    self.df_new = pd.read_csv(filename,header=None,
                              names=self.names, delimiter=',')

    self.df_new['date'] = pd.to_datetime(self.df_new['date'])
    self.df_new['daynum'] = self.df_new['date'].dt.dayofyear
    self.df_new['wdaynum'] = self.df_new['date'].dt.dayofweek
    self.df_new['month'] = self.df_new['date'].dt.month
    self.df_new['year'] = self.df_new['date'].dt.year
    self.df_new['ymd'] = pd.to_datetime(self.df_new['date']).dt.strftime('%m/%d/%Y')
    self.df_new['ydecimal']=self.df_new['year']+self.df_new['daynum']/365.25
    self.df_new.sort_values(by = 'ydecimal')

    self.df_new = self.df_new[self.df_new['year'].isin(yr_list)]
```

Figure 5-6. *These **atads_figures** methods create the **self.var_dict** dictionary, build a list of available airports by inventorying the files in the APT_CSV folder, and read in all the requested airport in the **apt_ list**. Each airport is read using the Pandas **read_apt()** method which also adds a new column where the date is expressed as a decimal year (**ydecimal**).*

The update_mainchart( ) Method

This is the main method used to create the graphical output. It relies on the curve fitting, line plotting, and chart enhancing utilities, as seen in Figure 5-7.

This method's arguments receive lists of airports and years and parameters that specify whether to include the raw curve, a smoothed version of the raw curve, or polynomial fits and which variable (dataframe column) to use. Note the data column being used is specified by the **active_variable** argument.

For each airport, its data is extracted from **self.df** into **self.df_apt** – used by the utilities that construct the actual charts.

The **if** statements optionally include the various traces, via our **add_airport_trace()**, **add_smooth_trace()**, and **add_poly_trace_and_eqnstr()** methods addressed later.

Note that **show_poly** can be either 0, 1, or 2; when 0, no curve fitting is done; when nonzero, either the linear or quadratic curve fitting is used.

To minimize clutter, curve fitting, and equation display will only be done for the first two airports (**apt_count** < 3).

After all airport curves have been added, chart enhancements such as borders, colors, watermarks, and titles are added and the x-axis range and mouseover activated.

```
def update_mainchart(self,airport_list, yr_list, active_variable, smoothing,show_raw,show_poly):

        var_str = self.var_dict.get(active_variable)
        self.eqn = ""
        title_str0=':'.join(airport_list)
        title_str0 = title_str0 + ' ['+ var_str+']'+" Traffic by Year "

        year_min = float(min(yr_list))
        year_max = float(max(yr_list)+1.)

        self.fig_main = go.Figure()
        apt_count = 0
        for apt in airport_list:
                apt_count = apt_count+1

                self.df_apt = self.df[self.df['facility'].isin({apt})]

                if show_raw == '1':
                    self.add_airport_trace(apt, active_variable)

                if smoothing != '0':
                    self.add_smooth_trace(apt, smoothing, active_variable)

                if show_poly != '0' and apt_count < 3:
                    p = int(show_poly)
                    self.add_poly_trace_and_eqnstr(p,apt,year_min, year_max,active_variable)


        self.add_watermark()
        self.fig_main.update(layout_xaxis_range = [year_min,year_max])
        self.add_titles(title_str0,active_variable)
        self.add_border()
        self.year_block_colors(year_min, year_max)
        self.fig_main.update_layout(hovermode='x unified')
```

Figure 5-7. *The **update_mainchart()** method is used to construct the prototype's chart from the user-specified parameters. Most of the **atads_figures'** other methods support this one method.*

Methods for Drawing Raw and Smoothed Data

There are two methods to support drawing the raw and smoothed data for the **active_variable** for a specified airport (see Figure 5-8a and an example using smoothing in Figure 5-8b). Both methods use the Plotly graphics object scatter() function previously encountered and add a new column to the dataframe (**self.df['smth']**).

add_airport_trace() adds the raw data to the chart and also sets the **hovertemplate** (mouseover) parameter information to the year-month-date and decimal year value which appears whenever the cursor is placed near a chart point.

add_smooth_trace() takes the smoothing number (i.e., the smoothing window size) and calculates the average value for the active variable, for each x-coordinate centered on the window. The x-coordinate is shifted by half a window width to correct an unwanted offset created from the **rolling()** function. By smoothing the data, short-term variations are suppressed, which visually emphasizes long-term patterns.

```
def add_airport_trace(self,apt,active_variable):
    self.fig_main.add_trace(go.Scatter(name=apt,
            x=self.df_apt['ydecimal'],
            y=self.df_apt[active_variable],
            connectgaps = False,
            text=self.df_apt['ymd'],
            hovertemplate=
            "<b>%{text}</b><br><br>" +
            " %{y}<br>" +
            "<extra></extra>"
            ))


def add_smooth_trace(self,apt,smoothing,active_variable):
    window=int(smoothing)
    self.df_apt['smth'] = self.df_apt[active_variable].rolling(window).mean()
    self.df_apt['smth'] = self.df_apt['smth'].shift(-window//2)
    self.fig_main.add_trace(go.Scatter(name=apt,
    x=self.df_apt['ydecimal'],
    y=self.df_apt['smth']))
```

Figure 5-8a. *Methods for plotting the raw and smoothed versions of the **active_variable** data*

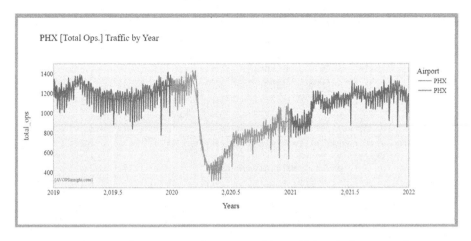

Figure 5-8b. *Applying a 30-day smoothing filter to the raw data for Phoenix (PHX) produces a curve (red) that nicely captures the seasonal changes. The raw data is shown as a blue curve.*

Methods to Enhance Chart Visual Appeal

These methods (see Figure 5-9) are similar to those introduced in Figure 2-2 and add borders, alternating background colors for each year, a watermark, and some titles. They are straightforward and self-explanatory applications of the PLOTLY libraries.

As previously noted, **paper** coordinates are used for the convenience of referring to the chart's region as extending from 0 to 1 in each axis, so annotations, for example, can be positioned without regard to data values or data scaling.

The watermark used is a reference to a demo website (avopsinsight. com) I created to test the software and for student use.

```
def add_watermark(self):
    self.fig_main.add_annotation(
            xref="paper",yref="paper",
            x=0, y=0,
            text="[AVOPSinsight.com]",
            font=dict(family="sans serif", size=10, color="LightSlateGray"),
            showarrow=False,
            yshift=10)

def year_block_colors(self,year_min, year_max):
        for y in range(int(year_min),int(year_max)):
            if (y % 2) == 0:
                    self.fig_main.add_vrect(
                        x0=y,
                        x1=y+1,
                        row="all",
                        col=1,
                        fillcolor="mistyrose",
                        opacity=0.4,
                        line_width=0)


def add_titles(self,title_str,active_variable):
    self.fig_main.update_layout(
        title=title_str,
        font=dict(family="sans serif", size=14, color="Blue"),
        xaxis_title="Years",
        yaxis_title=active_variable,
        legend_title="Airport",
    )

def add_border(self):
    self.fig_main.update_layout(shapes=[go.layout.Shape(
        type='rect',
        xref='paper', yref='paper',
        x0=0., y0=0.,x1=1.0, y1=1.0,
        opacity=.4,
        line={'width': 1, 'color': 'black'}
    )])
```

Figure 5-9. *The class methods used to make our charts more visually appealing by adding watermarks, alternating year colors, titles, and a border*

Methods to Add Polynomial Curve Fits

We don't need to master the mathematics of curve fitting. We can import the **poly** library to do the curve fitting: **poly.poly()** finds coefficients, while **poly.polyval()** uses coefficients to return a list of y-values matching the fitted curve.

There are two method groups to support polynomial curve fitting. The first (see Figure 5-10) uses **add_poly_trace_and_eqnstr()**, which extracts the needed data from the **self.df** dataframe, based on year range, airport, active variable, and polynomial order. It then calls **get_poly_coeffs()** to return the polynomial coefficients using the **poly.polyfit** library.

Using the coefficients, the polynomial is drawn using **draw_poly()** which invokes the **poly.polyfit()** function to return a set of y-values so the fitted curve can be plotted and an equation string (**self.eqn**) constructed similar to equations 5.1 or 5.2 and added to the chart by the **self.add_eqn_str()** method.

The second group of methods supporting our curve fitting are simply routines that can convert the coefficients returned by the polynomial fitting process into equations like Figure 5-1 or 5-2, suitable for inclusion onto our charts. These are shown in Figure 5-11.

There are multiple challenges being met here. First, we need to augment the equations with airport names, so the strings are more user-friendly. Second, we want to keep the mathematical +/– symbols elegant with appropriate numbers of decimal places; and third, we make a mathematical transformation, so our equations are for time intervals after the start year – otherwise, they would mathematically refer to 0 A.D.!

get_apt_linear_equation_string() and **get_apt_quadratic_equation_string()** are the methods used to create the displayed equation strings. They are similar except the latter is a little more involved because it involves a quadratic term.

Both methods take the input coefficients c[] and convert them into new coefficients c00, c11, and perhaps c22 to work with intervals after the starting year.

To understand what's happening here, consider the linear equation form, and remember that c[1] is the slope of the line, and c[0] is the intercept – the value when t = 0.

Assume an airport in the year 2000 A.D. had 800 daily operations, and the growth rate for that year was an additional 10 operations on average. This means the trend (slope) is c[1] = 10/yr. What is c[0]? c[0] must obey Equation 4.1, so if y = 800, and c[1] is 10, then 800 = c[0] + 10 × 2000, and so c[0] = 800 – 20000 = –19200.

```
def add_poly_trace_and_eqnstr(self,p,apt,year_min,year_max,active_variable):
    eqn_str=""

    self.df_vals = self.df_apt[self.df_apt['ydecimal'].between(year_min, year_max)]
    coefs = self.get_poly_coeffs(apt,p,year_min,year_max,active_variable)
    self.draw_poly(apt,coefs)
    if p == 1:
        eqn_str = eqn_str + self.get_apt_linear_equation_string(apt,coefs,year_min)

    if p == 2:
        eqn_str = eqn_str + self.get_apt_quadratic_equation_string(apt,coefs,year_min)

    self.eqn = self.eqn + eqn_str
    self.add_eqn_str()

def get_poly_coeffs(self,apt,poly_order,year_min,year_max,active_variable):
    p = int(poly_order)
    coefs = poly.polyfit(self.df_vals['ydecimal'].values, self.df_vals[active_variable].values,p)
    return coefs

def draw_poly(self,apt,coefs):
    ffit = poly.polyval(self.df_vals['ydecimal'], coefs)
    self.fig_main.add_trace(go.Scatter(name=apt,
        x=self.df_vals['ydecimal'],
        y=ffit))

def add_eqn_str(self):
    self.fig_main.add_annotation(
            text=self.eqn,
            xref="paper",yref="paper",
            x= 0, y=1.1,
            font=dict(
                family="sans serif",
                size=16,
                color="Black"
                ),
            showarrow=False
            )
```

Figure 5-10. *These methods use **get_poly_coeffs()** to get the coefficients needed by **draw_poly()** to add the curve fit polynomial to the chart and build and add the text equation strings to the chart.*

Let's test this. How many operations were there in the year 2000? We expect y should be 800, t is 2000, c[1] is 10, and c[0] is –19200, so

$$y = -19200 + 10 \times 2000 = 800$$

which works! But this is a little awkward, since calculating y for 2003 requires calculating $y = -19200 + 10 \times 2003$ and so on. We know it must be an additional 30 operations, three years after 2000, and so on; there must be an easier way.

If instead of using the time since 0 A.D. we used the time since 2000 A.D, our equation would be simpler: $y = 800 + 10 \times t$, where t is now the years after 2000. It's still a straight line; we are using new coefficients c00 = 800 and c11 = 10. For the linear model, the trend/slope/rate didn't change, but the intercept did.

It's a little more complicated for quadratics, but the point here is that new coefficients must, and can, be calculated from the originals (provided by the **poly_fit** routine), if we wish to refer to **year_min** instead of 0 A.D. for our time reference.

If all of this is a little confusing, just remember **poly_fit()** gives us coefficients (c[0], c[1], etc.) assuming we are measuring time from 0 A.D, but we are converting them to coefficients (**c00**, **c11**, etc.) that assume time is measured from the start of our time frame (**year_min**).

```
def get_apt_linear_equation_string(self,apt,c,year_min):
 c00 = c[0] + year_min*c[1]
 c11 = c[1]

 a00 = "{:.1f}".format(c00)
 a11 = "{:.1f}".format(c11)

 apt_str = " [" +apt +": Ops = "
 if c11 < 0:
     sign11 = ' - '
     a11="{:.4f}".format(-c11)
   # eqn_str = eqn_str + ' + '+ c11 + ' * t ]  '
 else:
     sign11 = ' + '

 apt_str = apt_str + a00 + sign11 +a11 + ' * t ] '
 return apt_str

def get_apt_quadratic_equation_string(self,apt,c,year_min):
     apt_str = " [" +apt +": Ops = "
     t0 = year_min
     c00 = c[0] + c[1]*t0 + c[2]*t0*t0
     c11 = c[1] + 2*c[2]*t0
     c22 = c[2]

     a00="{:.1f}".format(c00)
     a11="{:.1f}".format(c11)
     a22="{:.1f}".format(c22)

     if c11 < 0:
         sign11 = ' - '
         a11="{:.1f}".format(-c11)
     else:
         sign11 = ' + '

     if c22 < 0:
         sign22 = ' - '
         a22="{:.1f}".format(-c22)
     else:
         sign22 = ' + '

     apt_str = apt_str + a00 + sign11 +a11 + ' * t ' + sign22 + a22 + ' * t * t ]
     return apt_str
```

Figure 5-11. *Methods to convert the coefficients into a text string showing an equation*

The rest of the effort for these two methods is assembling the equation string with the proper sign and numeric formatting by creating a string version of each coefficient and a more elegant numeric sign with a space on both sides.

Fine-Tuning with CSS

While we can display our PLOTLY/DASH output in a web browser without using CSS, realistically, when working with more complicated dashboards, we can, and should, use CSS to control the widget and chart screen layout. Just like any other form of coding, writing CSS can become quite a chore; however, there are some basic techniques that can be implemented which, with a relatively small amount of effort, will give you considerable control over your dashboard's look and feel, and in keeping with our overall strategy, we are not trying to provide a comprehensive overview, but rather a concrete example to refer to.

Our goal here is to create a CSS file which PLOTLY will automatically read if it is at the same directory as our Python code, and we will use this file to control the layout (e.g., widget and chart sizes, positions, background colors, rounded corners). (Be careful here – other CSS files in embedded folders might also be detected and produce hard-to-track unwanted behaviors!) When we developed our **atads_layout** class, we used **ids** so callbacks could manage input and output flows, but we also included a **className** parameter in some of the **html.Div()** methods. CSS can use **className** values to position elements on our display.

Look again at Figure 5-1 which shows our dashboard. Notice there are two panels in a side-by-side layout; the panels are not at the top of the window – there is a gap above and below; there are colors set for backgrounds, and there are rounded corners; the chart and widgets have some padding, so they are nicely presented and not bunched up against enclosing rectangles. This was all done using CSS.

CSS is extremely powerful and gives experts an ability to fine-tune web page layout, but nuanced layouts and appearance can require extremely arcane commands and an unwanted level of effort and distraction since most engineers and researchers care more about getting the dashboard's contents to their end users than becoming web specialists. One solution is to use CSS grid displays which offer a reasonable approach to layout by

85

allowing us to define a grid overlay consisting of rows and columns. The rows and columns allow us to define individual or rectangularly grouped cells, and these can be used to hold charts and widgets.

The cell groupings (which I will call panels) give flexibility and can hold widget collections, charts, etc., of different sizes, so we are not restricted to the size of an individual grid cell, and in addition, we can control the spacing of the grid's rows and columns. Grids therefore allow us to add, and position, panels of different sizes (as long as they are rectangular) on our display.

Let's now examine the CSS file used to create Figure 5-1's display.

In our display, we have multiple rows, but only the second is used. The first row allows for a small logo and a banner to be added later. The second displays our parameter selection widgets and the main chart. There are (unused) rows for additional dashboard panels to be added later.

At the top of our CSS file, we have the CSS code for the **app.layout** assignment's **className='content'** statement in the atads.py file (see Figure 5-12). We specify we are using the grid layout, and we set a template for the rows and columns. There are six columns and are all equally wide, sharing the same fraction of the width. If I wanted one column to be twice as wide as any of the others, I would have specified 2fr for that column.

The grid row heights I specified explicitly – it was the easiest way to get started. The first row is 150px, the second is taller (480px), and so on. The remaining rows are placeholders for later development.

```
.content {
background: #555555;
display: grid;
grid-template-columns: 1fr 1fr 1fr 1fr 1fr 1fr;

grid-template-rows: 150px 480px 150px 480px 500px 200px;
grid-gap: 1%;
padding: 1%;
border-radius: 10px;
overflow: hidden;
}
```

Figure 5-12. *Part I of the CSS file sets up the CSS grid display, the number of rows and columns, and also their relative sizes.*

The first row's CSS is shown in Figure 5-13 – it is a placeholder for a banner to be installed later. We use the grid-row-start and grid-row-end (and similarly for columns) to define the screen display regions/panels assigned to the banner. The notation ".banner{}" refers to a CSS class defined in our code, and the code for a CSS class can be used to set many parameters, such as font sizes, alignment, and colors for that class. In this case, the banner CSS code block is simply a placeholder for when we add a banner over our dashboard in the next chapter. Being unspecified in the **atads.py** code means the first row of the display is empty. The grid display layout allows you to build in elements as you develop your dashboard and maintain placeholders for later use.

```
/* ROW 1 - Banner */

.banner {
  max-width: 100%;
  max-height: auto;
  grid-column-start: 1;
  grid-column-end: 7;
  grid-row-start: 1;
  grid-row-end: 2;
  background: #bab829;
  padding: 2%;
  color: #1111ff;
  align-content: center;
  font-size: 100%;
  border-radius: 10px;
  border:1px solid white;
}
```

Figure 5-13. *Part II of the CSS file controls the first row which has a placeholder for a logo on the left and a banner on the right. The allocated display space for each is determined by the row and column start and end parameters.*

It is worth noting that there is flexibility in how code assigned to classes is applied. For example, ".class1 .class2 {}" would apply the code in the curly brackets to class1 and class2. On the other hand, ".class1.class2 {}" would mean only apply the code when both class1 and class2 are present.

The second row's CSS is shown in Figure 5-14. Notice the **atads_layout** class used **parameter_selections** and **chart** as **className** settings in **html.Div()** statements, and our CSS is now referring to them when building this row's grid structure.

```
/* ROW 2 year/apt selection and time series chart */

.parameter_selections {
  background: #aaaaff;
  grid-column-start: 1;
  grid-column-end: 3;
  grid-row-start: 2;
  grid-row-end: 3;
  padding: 3%;
  padding-right: 5%;
  color: #444444;
  font-size: 105%;
  border-radius: 10px;
  border:1px solid white;
}

.chart{
 background: #aaaaff;|
  grid-column-start: 3;
  grid-column-end: 7;
  grid-row-start: 2;
  grid-row-end: 3;
  border-radius: 10px;
  border:1px solid white;
  overflow: hidden;
  padding: 1%;
}
```

Figure 5-14. *Part III of the CSS file is used to create the second row of our display grid. **.chart{}** refers to the **chart** class defined in an **html. Div()** statement in **atads_layout**.*

For visual appeal, I have added a blue background to the panels, rounded corners, and a white 1px wide border.

Summary

Our first draft of a dashboard has laid a foundation for enhancement since we have learned how to add panels to our dashboard. In the next chapter, we will build on these techniques to increase our dashboard's utility.

CHAPTER 6

Dashboard Enhancements

We have made considerable progress with our dashboard and learned how to create a reactive application that, being browser-friendly, could be accessed universally if we deploy it on a server. Before delving into the deployment in a server, a topic we will explore in the next chapter, we'll first make some improvements to enhance the appearance and functionality. Because we have used Python classes to organize our code, it will be easy to add new components by making local changes. By this, I mean that changes (additions and modifications) can be made with a reasonable assurance that they will not break something elsewhere.

The enhancements we wish to make will be to improve the appearance of the dashboard by adding a banner, information panels, and additional charts. Some new charts will present histograms showing operational counts by month and by day of the week, so, for example, questions concerning the busiest month or day of the week can be answered.

We will also add a Discrete Fourier Transform (spectrum) chart – an advanced technique used by scientists and engineers to identify periodic patterns in data – because this kind of technique can be very powerful with time series data; it will also allow us to demonstrate how easy it is to add sophisticated methods to a well-designed dashboard.

© Padraig Houlahan 2024
P. Houlahan, *Prototyping Python Dashboards for Scientists and Engineers*,
https://doi.org/10.1007/979-8-8688-0221-8_6

Very simply, a spectrum is a list of values where each value indicates how much of a certain frequency is present. If, as an input, I had a perfect (sine) wave of frequency 10 and obtained its spectrum spanning frequencies 0, 1, 2, 3, ... 100, then ignoring aliasing and other artifacts, this spectrum could be represented as a list of 101 entries, all being 0 except the 10th. In general, larger values in a spectrum indicate the degree to which frequencies are significant. For the record, frequency and period are reciprocals: a frequency of 1/10 corresponds to a period of 10. Frequencies are measured "per unit time," such as "once per day."

A little overkill perhaps for aviation data, but fascinating, nonetheless, as we shall see.

The final dashboard version we are building is shown in Figure 6-1. It has eight panels: Row 1 is a banner; Row 2 has the parameter selection and main chart panels; Row 3 is a general informational, a monthly and a weekday, panel; and on Row 4, there are panels for information and for spectrum.

We have already seen how to build Row 2, and we will use the same strategies to add all the other panels. The biggest differences will be that panels involving charts will require their own highly customized update methods, but other than for the update methods, the needed changes will be simple additions based on our existing panel entries.

All enhancements (new panels) will require adding new methods and **html.Div()** entries to the **atads_layout** class since we need the **html. Div(className=)** entries to create their panels with our CSS grid, and the CSS file will need code blocks to position and size them. The new elements will also need to be invoked from the **app.layout()** statement in **atads.py**. However, the charts, being dynamic, will also require callback entries and their own update functions and methods in the **atads_figures** class.

Figure 6-2 shows the relevant part of the **atads.py** file, where we can see how its **app.layout()**, callbacks, and updates were added. Remember, static panels like the banner and instructions do not need update or callback entries.

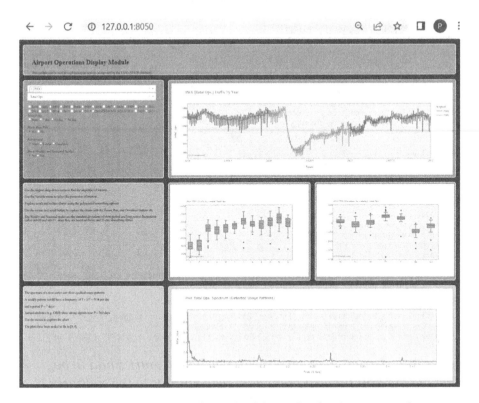

Figure 6-1. *The ATADS analysis dashboard. The top row is the banner, the second we have already completed. The third row's three panels are a general informational one followed by histograms showing monthly and day-of-week effects. The fourth row has brief information about the spectrum to the right, which in this case shows peaks that reveal weekly, semiweekly, and three times a week activity at Phoenix Sky Harbor. Pandemic effects are clearly shown in the main chart*

```
app.layout = html.Div(
            className="content",
            children=[
            configure_settings(),
            my_layout.chart(),
            my_layout.instructions(),
            my_layout.banner(),
            my_layout.mchart(),
            my_layout.wchart(),
            my_layout.spectrum_instructions(),
            my_layout.spectrum_chart()
            ])

@app.callback(
    Output("mchart",        "figure"),
    Output("wchart",        "figure"),
    Output("scatter_plot",  "figure"),
    Output("spectrum",      "figure"),
    Input("airports",       "value"),
    Input("years",          "value"),
    Input("use_variable",   "value"),
    Input("smoothing",      "value"),
    Input("show_raw",       "value"),
    Input("show_poly",      "value")
    )
def update_dashboard(airport_list, yr_list, active_variable, smoothing,show_raw,show_poly):
    my_figs.get_airport_data(airport_list,yr_list)
    my_figs.update_wchart(airport_list, yr_list, active_variable)
    my_figs.update_mchart(airport_list, yr_list, active_variable)
    my_figs.update_mainchart(airport_list, yr_list, active_variable, smoothing,show_raw,show_poly)
    my_figs.update_spectrum(airport_list, yr_list, active_variable)
    return  my_figs.fig_monthly, my_figs.fig_weekly, my_figs.fig_main, my_figs.fig_spectrum
```

Figure 6-2. *To add the new panels, new layout entries had to be added to the **app.layout()** function, new outputs to the callbacks, new updates to the **update_dashboard()** body, and new return values*

Adding the Banner and the Instruction Panels

We can make our dashboard more attractive to the end user by adding a banner across the top and by adding simple instruction panels, one for the parameter selection panel and another briefly explaining spectra. These are static elements and will not involve DASH callbacks, so are very similarly implemented.

To add a banner, we will add a new method to our **atads_layout** class called **banner()** which sets **className='banner'** that is invoked in the CSS file as a **.banner{}** block.

To add the instruction panels, we use the same strategy: create new **instructions()** and **spectrum_instructions()** methods in **atads_layout.py** that use **className** values of the **instructions** and **spectrum_instructions** panels. These new **className**s are then used in the CSS file by the **.instructions{}** and **.spectrum_instructions{}** blocks.

The CSS file blocks for these static panels are shown in Figure 6-3.

```css
.banner {
  max-width: 100%;
  max-height: auto;
  grid-column-start: 1;
  grid-column-end: 7;
  grid-row-start: 1;
  grid-row-end: 2;
  background: #bab829;
  padding: 2%;
  color: #1111ff;
  align-content: center;
  font-size: 100%;
  border-radius: 10px;
  border:1px solid white;
}

.instructions {
 background: #eeaaff;
  grid-column-start: 1;
  grid-column-end: 3;
  grid-row-start: 3;
  grid-row-end: 4;
  border-radius: 10px;
  border:1px solid white;
  overflow: hidden;
  padding: 1%;
}

.spectrum_instructions {
 background: #aaddbb;
  grid-column-start: 1;
  grid-column-end: 3;
  grid-row-start: 4;
  grid-row-end: 5;
  border-radius: 10px;
  border:1px solid white;
  overflow: hidden;
  padding: 1%;
}
```

Figure 6-3. *Using our CSS file's grid, we can position the banner and instruction panels using the .banner{}, .instructions{}, and the .spectrum_instructions{} blocks.*

Note, when you add new panels/elements to your dashboard, you might find the rendering doesn't quite work right unless you modify CSS grid column and row dimensions.

To actually add the banner and the instructions, their methods, **my_layout.banner()**, **my_layout.instructions()**, and **my_layout.spectrum_instructions()**, are added to the **app.layout** statement in the **atads.py** file (see Figure 6-4).

```
def banner(self):
    return      html.Div(
                    className="banner",
                    children=[
                    html.H1("Airport Operations Display Module "),
                    html.P("This module can be used to explore airport \
                            activity as reported by the FAA's ATADS database."),
                    ])

def instructions(self):
    return        html.Div(
          className="instructions",
          children=[
              html.P("Use the Airport drop-down menu to find the airport(s) of interest."),
              html.P("Use the Variable menu to select the parameter of interest. "),
              html.P("Explore trends and reduce clutter using the polynomial \
                        smoothing options"),
              html.P("Use the mouse and scroll button to explore the charts \
                        with the Zoom, Pan, and Download buttons etc. ")

          ])

def spectrum_instructions(self):
    return          html.Div(
          className="spectrum_instructions",
          children=[
              html.P("The spectrum of a time series can show cyclical usage patterns."),
              html.P("A weekly pattern would have a frequency of f = 1/7 = 0.14 per day"),
              html.P("and a period P = 7 days"),
              html.P("Annual airshows (e.g. OSH) show strong signals near P = 360 days"),
              html.P("Use the mouse to explore the chart."),
              html.P("The plots have been scaled to lie in [0,1]."),
              html.Br(),
              ])
```

Figure 6-4. *The new methods added to the **atads_layout** class. Each uses a **className** setting that uses a corresponding class block in the CSS file. Both elements are static but can be easily modified in these definitions for later dashboard versions.*

Monthly and Weekday Histogram Panels

The monthly and weekday charts require very similar changes to the dashboard, but they will have an important difference: to better visualize day-of-week effects, the time series will be smoothed with a 21-day window to remove seasonal effects, and the smoothed version subtracted from the original, to flatten it, leaving the weekday effects (mostly). A close-up of the dashboard showing these charts is shown in Figure 6-5.

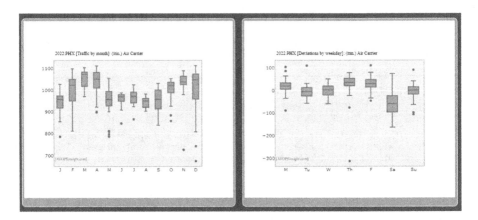

Figure 6-5. *The monthly and weekly histograms. Because the weekly is constructed by subtracting the local background in a 21-day moving window, it shows deviations from the background by day of the week.*

Because our dashboard will now have multiple charts/figures, we will also need to modify some existing code such as the **add_black_border()** method, which was originally written to use **self.fig_main**. To make it available for other charts, we will modify it slightly, so we can pass a figure as a function argument.

The **update_mchart()** method added to the **atads_figures** class is shown in Figure 6-6. It is intentionally restricted to the most recent year selected (because I felt it might be confusing if users combined multiple years since yearly trends can change substantially) and, like the main chart, adds a black border, titles, a watermark, and support for mouseover information. Unlike the main chart, it uses the graphics object **go.Box()** method to create the histogram.

```
def update_mchart(self,airport_list, yr_list, active_variable, smoothing,show_raw,show_poly):
        title_str0=':'.join(airport_list)

        self.fig_monthly = go.Figure()
        year_max = [max(yr_list)]
        year_str = str(year_max[0])+':'
        var_str = self.var_dict.get(active_variable)
        for i in airport_list:
                self.df_fac = self.df[self.df['facility'].isin({i})]
                df_fac_yr = self.df_fac[self.df_fac['year'].isin(year_max)]
                self.fig_monthly.add_trace(go.Box(name=i+':'+str(year_max),
                        x=df_fac_yr['month'],
                        y=df_fac_yr[active_variable]))

        self.add_watermark(self.fig_monthly,'[AVOPSinsight.com]')
        self.add_black_border(self.fig_monthly)

        self.fig_monthly.update_layout(
            xaxis = dict(
            tickmode = 'array',
            tickvals = [1, 2, 3, 4, 5, 6, 7, 8, 9, 10, 11, 12],
            ticktext = ['J','F', 'M', 'A', 'M', 'J','J','A','S','O','N','D']))

        self.fig_monthly.add_annotation(
            text=year_str+title_str0+" [Traffic by month]: " + var_str,
            xref="paper",yref="paper",
            x= 0, y=1.1,
            font=dict(
                family="sans serif",
                size=12,
                color="Black"),
            showarrow=False)

        self.fig_monthly.update_layout(hovermode='x unified', boxmode='group')
```

Figure 6-6. *The **update_mchart()** method is used to build the monthly histogram chart*

Notice how we changed the x-axis **tickmarks** to letters indicating the month by mapping the **ticktext** list to the numeric **tickvals**.

To create the weekday chart, we use a similar code but flatten the data, so seasonal changes do not affect the result. The method is called **update_wchart()** and is shown in Figure 6-7. Again, the main difference is that a smoothed version of the data is subtracted from the original to produce a flattened dataset. As we did for the monthly chart, the x-axis labels are converted from numeric to text using **ticktext/tickvals**. In doing this, the day-of-week counts are deviations from the longer-term trends and not absolutes.

```
def update_wchart(self,airport_list, yr_list, active_variable, smoothing,show_raw,show_poly):
    title_str0=':'.join(airport_list)
    self.fig_weekly = go.Figure()

    year_max = [max(yr_list)]
    year_str = str(year_max[0])+':'
    var_str = self.var_dict.get(active_variable)
    for i in airport_list:
        self.df_fac = self.df[self.df['facility'].isin({i})]
        df_fac_yr = self.df_fac[self.df_fac['year'].isin(year_max)]

        window = 21
        df_fac = df_fac_yr[active_variable]
        df_fac_smth = df_fac.rolling(window).mean()
        df_fac_smth = df_fac_smth.shift(-window//2)
        df_diff = df_fac - df_fac_smth
        self.fig_weekly.add_trace(go.Box(name=i+':'+str(year_max),
            x=df_fac_yr['wdaynum'],
            y = df_diff))

    self.fig_weekly.update_layout(
        xaxis = dict(
        tickmode = 'array',
        tickvals = [0, 1, 2, 3, 4, 5, 6],
        ticktext = ['M','Tu', 'W', 'Th', 'F', 'Sa','Su']))
    self.add_watermark(self.fig_weekly,'[AVOPSinsight.com]')

    self.fig_weekly.add_annotation(
        text=year_str+title_str0+" [Deviations by weekday]: "+var_str,
        xref="paper",yref="paper",
        x=0, y=1.1,
        font=dict(
            family="sans serif",size=12,color="Black"),
        showarrow=False )

    self.fig_weekly.update_xaxes(nticks=7)
    self.add_black_border(self.fig_weekly)
    self.fig_weekly.update_layout(hovermode='x unified',boxmode='group')
```

Figure 6-7. *The **update_wchart()** method creates the day-of-week histogram. It flattens the data by subtracting a smoothed version of the local data from itself (**df_diff**) to remove seasonal effects*

To handle the new charts, two new methods were added to the **atads_ layout** class, defining the IDs needed by the callbacks and the **className** settings needed by the CSS; see Figure 6-8.

```
def mchart(self):
    return              html.Div(
            className="monthly_chart",
            children=[
            dcc.Graph(id="mchart"),
        ])

def wchart(self):
    return              html.Div(
            className="weekly_chart",
            children=[
            dcc.Graph(id="wchart"),
        ])
```

Figure 6-8. *Methods added to the **atads_layout** class define the IDs needed by the callbacks and the **classNames** needed by the CSS file*

With these in place, the atads.py file uses the **app.layout statement** to call the new layout methods and additional outputs added to its **update_ dashboard()** callbacks and return values list. These changes can be seen in the atads.py segment shown in Figure 6-9.

The final piece is to add entries to the CSS file to control the grid and sizing.

```
.monthly_chart{
 background: #aaaaff;
  grid-column-start: 3;
  grid-column-end: 5;
  grid-row-start: 3;
  grid-row-end: 4;
  border-radius: 10px;
  border:1px solid white;
  overflow: hidden;
  padding: 1%;
}

.weekly_chart{
 background: #aaaaff;
  grid-column-start: 5;
  grid-column-end: 7;
  grid-row-start: 3;
  grid-row-end: 4;
  border-radius: 10px;
  border:1px solid white;
  overflow: hidden;
  padding: 1%;
}
```

Figure 6-9. *The CSS file blocks setting the grid parameters to position the new charts*

The Spectrum Panel

A spectrum panel chart is shown in Figure 6-10. Placing the cursor near a peak shows the associated frequency and period. Here, there is a peak at a frequency of 0.144/day which is 1/7 days – equivalent to once per week. There are other obvious peaks at two and three times this frequency, corresponding to twice and three times per week. A proper explanation of spectra is beyond the scope of this work, but qualitatively, we can see how they can reveal specific frequencies and periods in time series data and allow us to compare and explore airports.

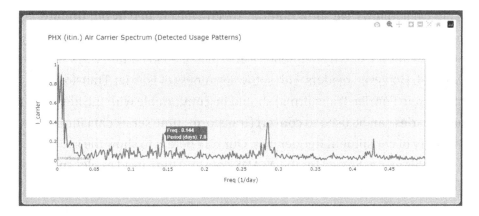

Figure 6-10. *The spectrum corresponding to a year of ATADS Itinerant PHX Air Carrier data (i_carrier)*

Just as for the other panels, the spectrum chart needs an entry in **atads_ layout()** and in the CSS file, shown in Figures 6-11a and 6-11b, respectively.

```python
def spectrum_chart(self):
    return  html.Div(
            className="spectrum_chart",
            children=[
            dcc.Graph(id="spectrum"),
        ])
```

Figure 6-11a. *The **atads_layout** class entry for the spectrum*

```css
.spectrum_chart{
 background: #aaddbb;
  grid-column-start: 3;
  grid-column-end: 7;
  grid-row-start: 4;
  grid-row-end: 5;
  border-radius: 10px;
  border:1px solid white;
  overflow: hidden;
  padding: 1%;
}
```

Figure 6-11b. *The CSS block used for positioning the spectrum chart*

In what follows, we show how our spectra are created by our **update_spectrum()** method. We provide a solution that works, but the reasons why the scales are set, or why frequencies are calculated in certain ways, are not provided. However, readers with some awareness of Fourier Transforms (or Discrete Fourier Transforms) should be comfortable with the idea that libraries can be used to convert (transform) time series data into an inventory of constituent frequencies. Our role here is to show the specifics of how this was achieved for the ATADS data and then used to plot spectra. The reader can copy and modify the algorithms for their own purposes.

Creating the actual spectrum uses many of the same strategies used for other charts where we essentially used an array of t[ ] and y[ ] data, where the t[ ] array was the dataframe's "**ydecimal**" column used for the x-axis, and the y[ ] array was the extracted values, for example, **df[active_variable]** used for the y-axis; DASH graphics objects were used to add the trace, add black borders, set the captions, etc.

For our spectrum, the x-axis is for frequencies, from zero up to some maximum determined by the input data's time range and the number of days being studied. The y-axis is for amplitude, a measure of how significant a frequency's contribution is. The general idea here is that, mathematically, we can consider a time series such as those shown in our dashboard's main chart, as consisting of a combination of waves or frequencies. As a simple example, a pure wave with a fixed wavelength (and hence frequency) could be equally represented by a single frequency and an amplitude (ignoring issues such as phase shifts). Such a wave would have a simple spectrum – a spike at the wave's frequency and zero everywhere else. The Discrete Fourier Transform is a mathematical tool to analyze such a sequence and provide a list of the amplitudes of each frequency present. To actually draw the spectrum, both the list of amplitudes and the list of corresponding frequencies are required.

Our **atads_figures** class' **update_spectrum()** method does this for us as part of building the spectrum (see Figure 6-12). It generates a rescaled array (**y_vals[ ]**) passed to the Fast Fourier Transform (FFT) library, and that outputs the amplitudes of the frequencies present (**a_vals[ ]**). The possible frequencies present are constrained by the number of input values and the day range and are calculated and stored in **fq_list[ ]**. The spectrum is simply a plot of the **a_vals** vs. the **fq_vals**.

There are only about a dozen lines of code specific to the problem of finding the amplitudes **a_vals[ ]** and frequencies **fq_list[ ]**. For each airport, its **active_variable** column is used to set the **y_vals[ ]** array from which the **mean()** value is subtracted. This is done to stop a spike appearing at zero frequency since a signal with zero frequency is a constant – related to the mean of the input data. By removing the mean value, we don't end up with a chart where a large peak at the origin sets a scale that hides smaller details elsewhere. The **fft()** function gives us a list of initial amplitudes, but we later rescale them, so their values range from 0 to 1, so we can overlay charts from diverse airports without rescaling. By its nature, the **fft()** function returns amplitudes assuming a specific set of frequencies (**fq_list[ ]**), and these are calculated separately using the number of data points and time interval. To allow the cursor to display both frequency and period, an array of periods is also calculated (**p_vals[ ]**).

Note the **fft()** function only receives the **y_vals[ ]** array – it receives no information concerning the actual times; it simply assumes the **y_vals[ ]** members are equally spaced in time. It is up to the programmer to build the corresponding frequency array, **fq_vals[ ]**.

Once we have the **a_vals[ ]** and **fq_list[ ]** arrays, we can proceed with our usual chart construction.

Since our time units are for daily airport operations counts, the frequencies are in "per day" or "1/day" units. For every frequency f, there is a period $p = 1/f$, so something with a weekly period $p = 7$ would have a corresponding frequency $f = 1/p = 1/7 = 0.144$. Both the frequency and the period are provided by the cursor. If something occurs twice a week, the frequency would be twice as large (0.244) with half the period (3.5 days).

```python
def update_spectrum(self,airport_list, yr_list, active_variable):
    title_str0=':'.join(airport_list)
    self.fig_spectrum = go.Figure()
    var_str = self.var_dict.get(active_variable)
    for i in airport_list:

        y0 = self.df[self.df['facility'].isin({i})]
        y0 = y0[active_variable] - y0[active_variable].mean()
        y_vals = y0                                 # set the input array for FFT

        a_vals = np.abs(np.fft.fft(y_vals))         # get the FFT amplitude array a_vals[]

        N = len(y_vals)                             # set scaling parameters and arrays
        n = np.arange(N)
        T = N
        freq = n/T
        n_oneside = N//2

        fq_list = freq[:n_oneside]                  # build list of frequencies
        a_vals = a_vals[:n_oneside]/n_oneside       # rescale amplitudes
        a_vals = a_vals /a_vals.max()               # Normalize to [0,1]

        fq_list[0] = 0.000001                       # avoid divide by zero
        self.p_vals = np.reciprocal(fq_list)        # build array of periodicities

        self.fig_spectrum.add_trace(go.Scatter(name=i,
            x=fq_list,
            y=a_vals,
            text=fq_list,
            customdata = self.p_vals,
              hovertemplate=
              "Freq.: <b>%{text:0.3f}</b><br>" +
              "Period (days): <b>%{customdata:0.1f}</b>" +
              "<extra></extra>") )

        self.add_watermark(self.fig_spectrum,'[AVOPSinsight.com]')

        self.add_black_border(self.fig_spectrum)
        self.fig_spectrum.update_layout(
            title=title_str0+ ' '+ var_str+" Spectrum (Detected Usage Patterns)",
            xaxis_title="Freq (1/day)",
            yaxis_title=active_variable,
            legend_title="Airport")
```

Figure 6-12. The **update_spectrum()** method uses NUMPY's **fft.fft()** to create a list of frequency amplitudes (**a_vals[ ]**). Various scaling parameters and arrays are used to create the frequency list (**fq_list[ ]**) and the corresponding list of periodicities (**p_vals[ ]**). The spectrum is a plot of the **a_vals** vs. the **fq_vals**.

Quantifying Weekly and Seasonal Effects

In this section, we address an interesting issue: Is there a way to quantify both the short-term (weekly) and long-term (seasonal/annual) variations often seen in our data? The challenge here is to figure out how to separate them so they can be estimated independently. As an example, we present a solution employing smoothing filters to mitigate smaller-scale variations in a data vector (specifically, a column in a dataframe). The outcome is a smoothed data vector that can be subtracted from the original, leaving behind a vector solely comprised of short-term fluctuations. This refined vector allows for the calculation of standard deviation, effectively documenting the scale of these fluctuations. On the other hand, choosing a sufficiently long smoothing filter window naturally results in a vector containing only long-term fluctuations, which can also be summarized with the standard deviation function.

In what follows, the discussion might be a little arcane for the nonspecialist, but it does demonstrate a very useful technique applicable to many datasets.

When viewing airport operations as a time series, we should be very aware of the fact that the data is not random in a Gaussian random noise sense and is often unbalanced or skewed. As an example, if we look at the main (time series) chart in Figure 6-1, there is an obvious asymmetry where the data shows noticeable drops weekly – there are no weekly spikes. There are also patterns relating to weekly cycles which are nonrandom, and there are seasonal trends, where there is often less traffic in the hotter months than in the cooler. We would like to be able to characterize these variations to better document trends and compare airports with another airport.

The quantity usually used in statistics to measure the spread of a set of values is the standard deviation, σ, where we might say a random variable with mean 20, and $\sigma = 2$, would be summarized with a statement like $x = 20 +/- 2$. This statement implies 99% of the actual values lie within

+/- 3σ of the mean and lie in the range 20-6 to 20+6 or between 14 and 26. Conversely, if we wanted to describe our data in terms of a signal or pattern's amplitude, if the signal's standard deviation was σ, the signal amplitude would be 3σ.

(Strictly speaking, it is a little problematic to interpret σ for our data in a purely statistical sense, because our data, being skewed, influenced by weekly cycles of human activity, and also seasonal weather cycles, does not normally present itself as a single-valued quantity with normal randomness. We *could* calculate σ for our data column, but we clearly would be losing information concerning the weekly and seasonal scales.)

Since our data appears to consist of two kinds of variations (signals), weekly and seasonal, we would like to isolate and measure them. The question now is this: How can we isolate the signals so that we can estimate their significance using either their standard deviations or amplitudes? (Note, this is not the same question as was addressed by our monthly and weekly histograms – those charts were based on averages taken over the year. To understand this, let's say a week's activity in January was [800, 850, 865, 865, 865, 865, 865], but in the heat of the summer, declined to [700, 750, 765, 765, 765, 765, 765]. Our weekly histogram would say the average activity was lowest for Mondays, around 775, but notice that in each week, the weekly variations range between 0 and 65. The variations, in January, are like a signal superimposed over an 800 seasonal background and similarly for the summer.)

Our solution will use smoothing filters (algorithms), which create new vectors from an original based on a rule that smears out the data over a specified timescale (the filter's "window"). As an example, if the filter window spans over five elements, then the value at position 87 in the new vector might be the average of elements located at positions 85, 86, 87, 88, and 89 in the original, that is, over the five elements centered on position 87 in the raw data. In this fashion, we can fill in all the elements in the new vector based on the old values. If we compute an average over five elements, variations will be smoothed out for scales shorter than five. In general, if we apply a smoothing filter of window size X, details with scales smaller than X are suppressed.

When applying a window-based filter, there will be boundary issues. For example, suppose we had a data column with 15 entries using a period 5 repeating pattern of the form

[101, 102, 103, 104, 105, 101, 102, 103, 104, 105, 101, 102, 103, 104, 105]

This pattern represents a period 5 sawtooth (triangular) wave ranging from 1 to 5, with a constant offset of 100.

If we apply a smoothing filter with a window of 5, then our smoothed data would look like

[NA, NA, 103, 103, 103, 103, 103, 103, 103, 103, 103, 103, 103, NA, NA]

where NA indicates Not Available. The NAs occur near the boundaries where the filter window wants to extend past the actual data. Because a filter window of size 5 always contains a combination of (101, 102, 103, 104, 105), the average will always be 103. In this example, the window size perfectly suppressed the original data's perfect sawtooth pattern. Also note that the first non-NA entry is at position 3, and we need to apply an offset of half the window size to align the smoothed data if there are other dataframe columns. [You can test this easily by commenting out the line in Figure 5-8 that uses the **shift()** function. If you do this, you will see the smoothed curves are offset from raw data.] If we now subtract our smoothed data from the original, we extract the triangular wave:

[NA, NA, 0, 1, 2, -2, -1, 0, 1, 2, -2, -1, 0, 1, 2, -2, -1, 0, NA, NA]

In this form, we see a sawtooth of mean 0 and amplitude 2. Our smoothed curve, subtracted from the original, allowed us to isolate signals with periods smaller than the filter window size.

As previously noted, very often, our data appears to consist of two patterns, one a weekly cycle and one showing long-term, seasonal changes that we might model as the sum of two vectors: $Y[\,] = Y_W[\,] + Y_S[\,]$. If we apply a 31-day smoothing filter to Y(), we would suppress the weekly variation, resulting in a vector $S_{31}[\,]$, a data sequence providing us with an estimate of the underlying Y_S similar to the smoothed curve seen in Figure 5-8. If we now use 9-day smoothing to suppress weekly effects, resulting in S_{09}, then subtracting S_{09} from Y would allow us to isolate the weekly variations: $Y_W = Y - S_{09}$. As an example, in Figure 6-13 we show how a 30-day filter was used to create the smoothed version of the raw data. The red curve would be S_{30} (30-day smoothed) vector, the blue one, and the Y vector. With 30-day smoothing, we have a curve that serves to provide a good estimate of long-term trends in the data. For this airport (PHX), there are very distinct weekly (short-term) variations, linked to airline activity which is generally more scheduled than small aircraft operations.

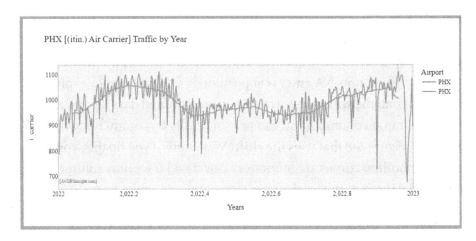

Figure 6-13. *Here, the red curve represents the results of applying a 30-day smoothing filter to the raw (blue) data; short-term information has been lost in creating it. However, subtracting the red from the blue would flatten the blue curve leaving short-term fluctuations.*

To summarize, we are viewing our data as having two main components: a changing seasonal background (red) in Figure 6-13 with weekly variations superimposed. To isolate weekly activity signals, subtract the 9-day smoothed data from the raw data. To isolate seasonal (monthly) variations, use 31-day smoothed data.

Now that we can isolate the short- and long-term signals into dataframe columns, we would like to characterize their significance. We will calculate their standard deviations and remember the signal amplitude is simply three times the standard deviation.

A word of caution: In our charts, we will refer to these two standard deviations calculated for the weekly and seasonal scales as **stdv09** and **stdv31**, respectively; we are deliberately avoiding the temptation to label them in the chart subtitles as weekly and monthly, for two reasons. First, we made arbitrary choices of 9 and 31 for our smoothing filter windows, and we wish to be explicit and force the user to remember our filter window sizes; another researcher might prefer to use 7 and 28 perhaps. Our choices were intended to avoid introducing artifacts into our analysis by using window sizes which are multiples of a week – a common fundamental resonance in the data. Second, while in this work, short-term, weekly, and **stdv09** all refer to the same scales, in using **stdv09** in the charts, my results will not be confused with a researcher's who decided to use 11 instead of 9 for the window scaling and who should then report their customized dashboard version's results using stdv11, etc.

But didn't we say that σ is used to measure things like Gaussian noise, and if so, why are we using it here? The answer is we are using σ because it is a known mathematical quantity and will measure the scale of variations to whatever data sequence we apply it to. The resulting number might have a slightly ambiguous statistical interpretation because of skewed data, but we are not making a statistical argument; we are simply using the standard deviation as a tool to estimate activity scales, so we can track trends and compare airports.

Now that we have a strategy for isolating and estimating the influence of weekly and seasonal patterns in our data, we would like to add this capability to our dashboard; we will need to modify the **atads.py** file to track a radio button toggle set to show our new scale estimates, **stdv09** and **stdv31**; modify our **atads_figures.py** file where charts are created; and add a radio button **radio_show_scales()** item to **atads_layout.py**.

The elements needed to add this capability to our dashboard's **atads. py** file are shown in Figure 6-14, while the **atads_figure.py** changes are shown in Figure 6-15 and **atads_layout.py** in Figure 6-16.

```
...

def configure_settings():
    return html.Div(
        className="parameter_selections",
        children=
        [

            ...
            html.Label('Show Scales'),                # add support for show_scales radio button
            my_layout.radio_show_scales()
        ])

...

@app.callback(
    ...
    Input("show_scales",        "value")            # add callback for show_scales
    )
def update_dashboard(...,show_scales):              # add show_scales to function arguments
    ...
    my_figs.update_mainchart(...,show_scales)       # pass show_scales to graph creating routine
    ...

...
```

Figure 6-14. *The **atads.py** file modifications needed to support showing the chart scales. We added an entry to support a new radio button to the parameter selection panel, a new callback Input() entry, and the **show_scales** parameter to the end of the update functions.*

In Figure 6-15, the method used to extract the scales is **get_scales()**, which for filter windows 9 and 31 adds two columns to the dataframe holding the smoothed S_{09} and S_{31} data. Then the **stdv09** estimate for the short-term signals is calculated from the original data minus S_{09}, while **stdv31** is calculated from S_{31}. **get_scales()** returns a string to be displayed as a chart subtitle.

```
...
  def get_scales(self,apt,y0,y1,active_variable):
      window=10
      self.df_fac['s09'] = self.df_fac[active_variable].rolling(window).mean()
      self.df_fac['s09'] = self.df_fac['s09'].shift(-window//2)
      window=30
      self.df_fac['s31'] = self.df_fac[active_variable].rolling(window).mean()
      self.df_fac['s31'] = self.df_fac['s31'].shift(-window//2)

      stdv09 = (self.df_fac[active_variable] - self.df_fac['s09']).std()
      stdv31 = (self.df_fac['s31']).std()
      str09 = "{:.1f}".format(stdv09)
      str31 = "{:.1f}".format(stdv31)
      stat_str = " [" +apt +": STDV09= " + str09 + '   STDV31= '  + str31 + ']'

      return stat_str

...
  def update_mainchart(...,show_scales):                    # add the show_scales parameter

          ...
          for i in airport_list:
              ...

              if show_scales == '1':
                  eqn_str = eqn_str + self.get_scales(i,year_min, year_max,active_variable)
                  self.eqn = self.eqn + eqn_str
          ...
```

Figure 6-15. *To calculate the short-term scale, stdv10, the standard deviation is calculated using the difference between the raw data (**df_fac[active_variable]**) and its 9-day smoothed version **s09**. **stdev31** is simply based on the 31-day smoothed version, **s31**.*

Finally, the **atads_layout.py** file has a new method to support the show scales radio button (see Figure 6-16).

```
...
    def radio_show_scales(self):
        return      dcc.RadioItems(id='show_scales',
                    options=[
                            {'label': 'No', 'value': '0'},
                            {'label': 'Yes', 'value': '1'}

                    ],
                    value='0',
                    labelStyle={'display': 'inline-block'}
                )
```

Figure 6-16. *To support the new radio button, a new method **radio_
show_scales()** is added to the **atads_layout.py** file.*

With these changes, row 2 of our dashboard looks like that in
Figure 6-17 where ANC and JFK are compared with the Show Scales
feature active for the year 2019.

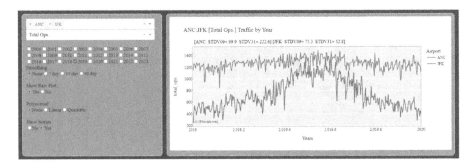

Figure 6-17. *ANC and JFK Air Carrier traffic for the year 2019 with
the Show Scales button active*

Even casual observation reveals that ANC's short-term or weekly
variations are more noticeable than those of JFK, aligning with their
respective standard deviation scales of 99 and 75. However, over the
year, JFK is much flatter than ANC; with STDV31 scales of 52 and 222,
respectively, a factor of four!

The Final ATADS Dashboard

With these changes, we now have a fully functional dashboard that allows us to compare multiple airports for different kinds of operations; explore data trends and find mathematical equations; and we can see monthly and weekday behaviors. The charts respond to mouseovers and have the built-in DASH capability to allow them to be selected, zoomed, and downloaded. We have also demonstrated how we can add new charts and panels if we wish, so our solution is scalable.

Is there such a thing as a completed dashboard project? I think the answer is both yes and no. If the dashboard fulfills its initial objectives, then yes; however, end users will almost certainly envision new features they would like to see implemented, so a successful dashboard will likely evolve.

In our case, there is still much that could be done. For example, we could consider splitting our **atads_figures** class into subclasses, and there are other kinds of charts we might like to add based on region and state. I elected to show multiple charts on the dashboard because there have been times when I noted a curious aspect in passing. As an example, while testing code I noticed some airports have significantly less day-of-week variation compared to others (i.e., PHX vs. JFK) that I might not have noticed if only the time series was displayed. However, having a single chart would be a possibility and could be implemented by having a set of radio buttons selecting different chart types and then using an if condition to select the one to be returned by the main callback.

Another improvement might be to use airport long names for pull-down menus instead of the three-letter versions. This would require creating additional lists and dictionaries similar to those done for the variable names. It might also be useful to add a feature where a chart's data could be displayed as a table and downloaded.

(You might have wondered about our CSS entries having redundancy such as the entries that create the rounded corners on our panels (**border-radius: 10px**) appearing in multiple elements because we simply copied and modified some blocks to create others. Should these not be cleaned up? Could we not put common elements into a block that is universally accessible? Maybe. There is no guarantee different browsers will interpret your CSS code the way you wish; in my experience, CSS is not as consistently implemented as a programming code like Python, and for small projects, redundancy gives stability and control. With a larger project, managed by a team, a member could delve into CSS nuances to organize the CSS, but for us, we will accept a little redundancy, so we can maintain a consistent working style and stay focused on more interesting issues.)

So, a project such as this will naturally suggest improvements with use and familiarity. With careful planning, it can avoid becoming ponderous if functionality is managed with a good divide-and-conquer design – using tabs on your browser display opens many possibilities for showing the end user smaller facets of a large project. Changes and improvements keep the dashboard relevant, and if well designed, these changes should be easy to make and give the developer a well-deserved sense of achievement.

Summary

In this chapter, we finished the atads dashboard by adding histograms to track seasonal and weekly effects and a spectrum chart. For aviation data, the spectrum is a novel or fun feature, but the underlying capability is very powerful if applied to other datasets.

We will next tackle the problem of how to deploy our dashboard, since we need a way for colleagues to access our work. We will use Unix server technology, which is well suited to this task and widely used by academic and industry researchers.

CHAPTER 7

Hosting an Application on a UNIX Server

Now that you have a working dashboard on your personal computer/ laptop, the next challenge is to make it available to your colleagues, and we will explore how this can be done using an Ubuntu (UNIX)-based server. In what follows, we will assume you will refer to some of the open source tutorials on the Internet which offer very detailed instructions on how to do most of what we need; the instructions can change if there are significant operating system changes, so be sure to use the appropriate versions.

We will need to do the following:

- Set up a Python virtual environment in which our software will be stored.

- Test the software with FLASK and uWSGI.

- Configure GUNICORN.

- Install NGINX.

- Create the virtual hosts in system configuration.

- Add a new service to the operating system.

- Secure our server with Fail2ban.

© Padraig Houlahan 2024
P. Houlahan, *Prototyping Python Dashboards for Scientists and Engineers*,
https://doi.org/10.1007/979-8-8688-0221-8_7

By creating a virtual environment, we can isolate our code from other software and configurations on our server. We could create separate virtual environments for different projects, and changes done in one virtual environment will not interfere with others.

But what about FLASK, WSGI, GUNICORN, and NGINX? What are they and why do we need them? With these, we are really entering the world of servers, and the terminology can be confusing, so let's clarify what these are.

WSGI (Web Server Gateway Interface) is a specification, a way of doing things, that allows web servers like APACHE and NGINX to work with Python applications; NGINX cannot talk directly to a Python application.

FLASK is a web framework, a WSGI application; it is not a server! FLASK has some limitations; being single-threaded, it cannot handle multiple users.

GUNICORN is a WSGI server. It can manage multiple worker processes to service interactions with your FLASK application. Being very efficient, it is well suited to this role, but it is not well suited for a public-facing server.

uWSGI is (very confusingly named!) a WSGI server, as is GUNICORN.

NGINX is a web server that can handle multiple clients, but it cannot communicate directly with a Python application. It can serve static pages but will send requests for dynamic pages to GUNICORN (our WSGI server of choice), which will translate the requests to WSGI to be used by FLASK.

In the world of servers, NGINX and GUNICORN are a very powerful combination.

We've established a chain of components connecting our Python application to the external-facing NGINX web server. Now, let's delve into the configurations of these different components.

We will build a simple Hello World web application to demonstrate the various steps. While many of the online recipes for doing something like this are terrific, when first learning, they can be confusing for many reasons. First, they often use app.py for the application, invoke the app as a flask instance inside app.py, and then pass **app:app** as a function parameter – that's a lot of apps! Second, when learning from examples, some recipes

will use **app.run()**, and others **app.run(host='0.0.0.0')** in their app.py
file with very different consequences we shall demonstrate. Third, some
demonstrations use a **wsgi.py** file that some servers relate to, but it's not
needed in other modes of running our app, and fourth, the **wsgi.py**
file is very similar to our **hello.py** file – a very confusing redundancy.

Creating the Python Environment

We want our project to be isolated, and therefore we will set up a virtual
environment to contain it. Virtual environments are mechanisms that
allow us to isolate projects from each other and can hold their versions
of libraries. We will use the Python **venv** package which must first be
installed along with some other useful packages in the usual fashion;
then we will create a virtual environment for our Hello World project
named **hwenv**, and finally we will activate it with the source command
(see Figure 7-1a). When the virtual environment is active, our prompt will
usually have something like **(hwenv)** embedded in it to remind us of which
environment is active. deactivate is a command used to exit from a virtual
environment. Note that the new environment will be a folder with the
name **hwenv**, and we will create our application files within that directory.

```
sudo apt-get update
sudo apt install python3-dev python3-pip python3-venv


python3 -m venv hwenv             # Create a virtual environment called hwenv

source hwenv/bin/activate
(hwenv)pip install flask gunicorn
(hwenv)pip install wheel
```

Figure 7-1a. *Installing the **venv** package, creating our environment
(**hwenv**) for our project, activating the environment, and installing
other packages (**wheel**, **flask**, and **gunicorn**) we will use later; **wheel**
is a package that helps ensure you can install software in your virtual
environment.*

The folder's contents can be seen using the UNIX **tree** utility (configured here to show three levels deep): **tree -L 3 hwenv**, the output of which is shown in Figure 7-1b where we can see local copies of various installed applications have been kept for our use. Also, if you explore the **activate** script, you will see it resets our home directory and search path which is how it really creates a private environment for the project and prevents us from accidentally loading other software versions residing on the server.

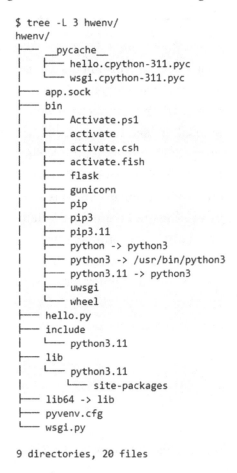

```
$ tree -L 3 hwenv/
hwenv/
├── __pycache__
│   ├── hello.cpython-311.pyc
│   └── wsgi.cpython-311.pyc
├── app.sock
├── bin
│   ├── Activate.ps1
│   ├── activate
│   ├── activate.csh
│   ├── activate.fish
│   ├── flask
│   ├── gunicorn
│   ├── pip
│   ├── pip3
│   ├── pip3.11
│   ├── python -> python3
│   ├── python3 -> /usr/bin/python3
│   ├── python3.11 -> python3
│   ├── uwsgi
│   └── wheel
├── hello.py
├── include
│   └── python3.11
├── lib
│   └── python3.11
│       └── site-packages
├── lib64 -> lib
├── pyvenv.cfg
└── wsgi.py

9 directories, 20 files
```

Figure 7-1b. *The output of the **tree** command after being applied to the **hwenv** folder shows how our Python environment keeps local copies of critical software. Running the **activate** command ensures only these copies will be used.*

Running a Flask Application

In the hwenv directory, create a file named **hello.py** (see Figure 7-2). In **hello.py**, we import the **flask** library and create an **app** – an instance of **Flask**. We use a decorator to set the path "/" and add a simple function to display a string "Hello World!" and a mechanism to run the app instance.

```
from flask import Flask
app = Flask(__name__)

@app.route('/')

def hello():
    return("Hello World!")

if __name__ == '__main__':
    app.run()
```

Figure 7-2. *Our flask application: hello.py*

It's a Python application, so let's test it (see Figure 7-3).

```
$ python hello.py
 * Serving Flask app 'hello'
 * Debug mode: off
WARNING: This is a development server. Do not use it in a production deployment.
    Use a production WSGI server instead.
 * Running on http://127.0.0.1:5000
Press CTRL+C to quit
```

Figure 7-3. *Running the first version of hello.py using Python*

It seems to work, but there is a catch that can be very annoying if not understood. Notice how the output suggests we should be able to see output at port 5000. But this is for the localhost 127.0.0.1. If you are running the application on an external machine instead of localhost, then you need to use the IP address of the external machine.

To fix this, and to turn on debugging while we're at it, change **app. run()** to **app.run(debug=True, host='0.0.0.0')**, which allows for all IPs (0.0.0.0), and rerun (see Figure 7-4).

119

```
$ python hello.py
 * Serving Flask app 'hello'
 * Debug mode: on
WARNING: This is a development server. Do not use it in a production deployment.
            Use a production WSGI server instead.
 * Running on all addresses (0.0.0.0)
 * Running on http://127.0.0.1:5000
 * Running on http://146.190.48.240:5000
Press CTRL+C to quit
 * Restarting with stat
 * Debugger is active!
 * Debugger PIN: 250-959-470
```

Figure 7-4. *Running the application with **app.run(debug=True, host='0.0.0.0')** specified. Notice how both the localhost 127.0.0.1 and our public-facing IP use port 5000.*

Our web page can now be seen (see Figure 7-5).

Figure 7-5. *The **hello.py** output can now be seen from remote sites.*

(If you encounter problems, make sure port 5000 is not being used by another project, and if not open, do "**sudo ufw allow 5000**" to open it through the firewall.)

We have seen how Python can use flask to send information to port 5000. What about **flask** itself? In Figure 7-6, I invoke **flask** and tell it to find the app instantiation in **hello.py** (**--app hello**), set the host parameter to work with all IPs (**--host=0.0.0.0**), and change the port number (**--port=5001**).

```
$ flask --app hello run --host=0.0.0.0 --port=5001
 * Serving Flask app 'hello'
 * Debug mode: off
WARNING: This is a development server. Do not use it in a production deployment.
         Use a production WSGI server instead.
 * Running on all addresses (0.0.0.0)
 * Running on http://127.0.0.1:5001
 * Running on http://146.190.48.240:5001
Press CTRL+C to quit
```

Figure 7-6. *Using flask to run the hello.py application, sharing it universally, and setting the port number to 5001*

If we go to port 5001, we will see the same output as in Figure 7-5.

So, we see that FLASK can also run a Python program and send the output to a UNIX port.

Using uWSGI

uWSGI is a WSGI server we can use to serve our Python application. Use "**pip install uwsgi**" to install it.

We can create a configuration file (**wsgi.py**) for uWSGI to find relevant information about our project (see Figure 7-7).

```
$ cat wsgi.py
from hello import app

if __name__ == "__main__":
    app.run()
```

Figure 7-7. *The wsgi.py file is the project's WSGI entry point. It holds all the information uWSGI needs, in this case to use the instance of app found in hello.py.*

This file simply says to use the **app** instance found in the **hello.py** file. When calling **uwsgi**, the entry point is passed in as a parameter **wsgi:app**. This is equivalent to "from wsgi.py import app". We will select HTTP as the communication protocol and keep 5000 as our port. Note, just as for our flask application's parameters, the 0.0.0.0 is not enclosed in single quotes. The command and some of the screen output are shown in Figure 7-8.

121

```
$ uwsgi --socket 0.0.0.0:5000 --protocol=http -w wsgi:app
*** Starting uWSGI 2.0.22 (64bit) on [Sat Sep 30 18:59:47 2023] ***
compiled with version: 12.3.0 on 30 September 2023 18:48:21
os: Linux-6.2.0-32-generic #32-Ubuntu SMP PREEMPT_DYNAMIC Mon Aug 14 10:03:50 UTC 2023
nodename: devsrv
machine: x86_64
clock source: unix
detected number of CPU cores: 1
current working directory: /home/pjh/hwenv
detected binary path: /home/pjh/hwenv/bin/uwsgi
!!! no internal routing support, rebuild with pcre support !!!
*** WARNING: you are running uWSGI without its master process manager ***
your processes number limit is 3657
your memory page size is 4096 bytes
detected max file descriptor number: 1024
lock engine: pthread robust mutexes
thunder lock: disabled (you can enable it with --thunder-lock)
uwsgi socket 0 bound to TCP address 0.0.0.0:5000 fd 3
...
```

Figure 7-8. *The result of running **uwsgi** where the HTTP protocol is specified. The entry point argument (-w **wsgi:app**) could be read as "from **wsgi.py** import **app**".*

While the **uwsgi** command is running, our web browser will again see "Hello World!" at port 5000 of the server.

Using GUNICORN

We have shown how our app can be run by **python**, **flask**, and **uwsgi**. Let's see how to run it using **gunicorn**. First, make sure **gunicorn** is installed, and it can be run as shown in Figure 7-9.

```
$ gunicorn --bind 0.0.0.0:5000 wsgi:app
[2023-10-01 22:06:15 +0000] [38446] [INFO] Starting gunicorn 21.2.0
[2023-10-01 22:06:15 +0000] [38446] [INFO] Listening at: http://0.0.0.0:5000 (38
[2023-10-01 22:06:15 +0000] [38446] [INFO] Using worker: sync
[2023-10-01 22:06:15 +0000] [38447] [INFO] Booting worker with pid: 38447
[2023-10-01 22:06:21 +0000] [38446] [INFO] Handling signal: winch
```

Figure 7-9. *The terminal output from executing the **gunicorn** command. The command successfully serves port 5000.*

The command line should be very intuitive to us now; **gunicorn** is asked to bind to port 5000 of all IPs and to work with the app instance referred in **wsgi.py**.

Note, while it is nice to use a **wsgi.py** entry point file, it should come as no surprise that the **gunicorn** command in Figure 7-9 also works if **hello:app** is used instead of **wsgi:app**.

Summary

We have come a long way with this chapter. We have taken our dashboard from our laptop and have added it to a Unix environment, which means it can now be shared and is in an environment that is stable and independent of the developer and hence is better placed to support team projects.

We still need to add other capabilities, such as to make the dashboard a unix service, so, for example, it will automatically restart when the server reboots, and to build support for it using a powerful web server such as Nginx. And there are issues of security – how to protect your server from malevolent actors. These issues will be addressed in the next chapter.

## CHAPTER 8

Deploying Your Project As a UNIX Service

Developing and running an application in a local machine and serving the application as a service on a server present an interesting challenge. At this point, we can run our application, and colleagues can access our dashboard remotely. But we should harden our design. Wouldn't it be nice if the software started automatically in the event of a server reboot? Without us having to log in and start the virtual environment each time?

At the same time, there might be many other projects and static web pages that need to be served, so you will likely have to either use an existing web server such as NGINX or APACHE or install them yourself.

Note that much of what we do in this chapter requires UNIX admin (**sudo**) privileges since we are working at levels above the normal users.

We have used no web servers until now – our **gunicorn or flask or uwsgi** applications simply provided the data to port 5000, and your web browser did the rest – it used that data to render the web page. To better manage our outward-faced services, we will use NGINX.

© Padraig Houlahan 2024

P. Houlahan, *Prototyping Python Dashboards for Scientists and Engineers*,
https://doi.org/10.1007/979-8-8688-0221-8_8

NGINX is a powerful and efficient web server and, when placed before GUNICORN, is called a "reverse proxy." It is easily installed using "sudo apt install nginx." You can build out your server's web presence by adding pages to the /var/www directory, so remote visitors can find information and links to your project.

NGINX will be a service offered by our server. Services are applications that are usually started automatically and are available whenever the server is rebooted that handle important server functions – anything from handling DNS requests to file management to user logins. They are usually multi-threaded – meaning they can handle multiple simultaneous requests. In going from the user to multiuser world, we move from applications to services.

When standard packages like NGINX or APACHE are installed, the installation process will configure all that is needed for them to act as services. In our case, we must build the service configurations for customized projects like ours.

Once a service is configured, it can be managed using the **systemctl** command which has the structure:

sudo systemctl action servicename

where we are using **sudo** to gain admin privileges.

There are five kinds of actions we most care about: **start/stop** for starting the service; **enable/disable** to cause it to start automatically on reboot; and **status**, which tells us about the service, whether it is running or enabled, and more.

In the rest of this chapter, we will first go through the processes of creating a service for our hello.py app and show how to share it with NGINX, before we show how to similarly process our atads project. The reason we are taking this approach is our atads project uses DASH and will necessitate some customizations not required by our non-DASH hello.py application.

Creating a Hello World System Service

An Ubuntu service can be created by adding a configuration file to the **/etc/systemd/system** directory. We will add a file called **hwapp.service** to support our hello-world app (hwapp) – see Figure 8-1.

```
[Unit]

Description=Gunicorn instance for hwapp project
After=network.target

[Service]
User=pjh

Group=www-data

WorkingDirectory=/home/pjh/hwenv
Environment="PATH=/home/pjh/hwenv/bin"

#ExecStart=/home/pjh/hwenv/bin/gunicorn --workers 3 --bind unix:app.sock -m 007 wsgi:app
ExecStart=/home/pjh/hwenv/bin/gunicorn --workers 3 --bind 0.0.0.0:5000 -m 007 wsgi:app

[Install]
WantedBy=multi-user.target
```

Figure 8-1. *The **hwapp.service** file which allows **hwapp** to act as a service - in this case uses the virtual environment in my home directory (pjh)*

The service configuration file contents are laid out in blocks.

In the **[Unit]** block, a description is provided and also a directive to wait until networking has first been established.

In the **[Service]** block, user and group memberships are set, along with information such as what directory to use. Essentially, the gunicorn command is used to start the **hwapp.py** application in our hwapp environment's directory. For now, we are using it to start services on port 5000 for all IPs, but because we will want to run the app as a Unix socket later, I include the Unix socket form as a comment for later.

Finally, the **[Install]** block indicates the Unix server needs to be in a multiuser state.

We can test our service with **sudo systemctl start hwapp.service** and check the results using **sudo systemctl status hwapp** – see Figure 8-2.

```
$ sudo systemctl status hwapp
● hwapp.service - Gunicorn instance for hwapp project
     Loaded: loaded (/etc/systemd/system/hwapp.service; disabled; preset: enabled)
     Active: active (running) since Wed 2023-10-04 18:16:54 UTC; 5s ago
   Main PID: 84601 (gunicorn)
      Tasks: 4 (limit: 1097)
     Memory: 60.9M
        CPU: 547ms
     CGroup: /system.slice/hwapp.service
             ├─84601 /home/pjh/hwenv/bin/python3 /home/pjh/hwenv/bin/gunicorn --workers 3 --bind 0.0.0.0:5000 -m 007 wsgi:app
             ├─84602 /home/pjh/hwenv/bin/python3 /home/pjh/hwenv/bin/gunicorn --workers 3 --bind 0.0.0.0:5000 -m 007 wsgi:app
             ├─84603 /home/pjh/hwenv/bin/python3 /home/pjh/hwenv/bin/gunicorn --workers 3 --bind 0.0.0.0:5000 -m 007 wsgi:app
             └─84604 /home/pjh/hwenv/bin/python3 /home/pjh/hwenv/bin/gunicorn --workers 3 --bind 0.0.0.0:5000 -m 007 wsgi:app

Oct 04 18:16:54 devsrv systemd[1]: Started hwapp.service - Gunicorn instance for hwapp project.
Oct 04 18:16:54 devsrv gunicorn[84601]: [2023-10-04 18:16:54 +0000] [84601] [INFO] Starting gunicorn 21.2.0
Oct 04 18:16:54 devsrv gunicorn[84601]: [2023-10-04 18:16:54 +0000] [84601] [INFO] Listening at: http://0.0.0.0:5000 (84601)
Oct 04 18:16:54 devsrv gunicorn[84601]: [2023-10-04 18:16:54 +0000] [84601] [INFO] Using worker: sync
Oct 04 18:16:54 devsrv gunicorn[84602]: [2023-10-04 18:16:54 +0000] [84602] [INFO] Booting worker with pid: 84602
Oct 04 18:16:54 devsrv gunicorn[84603]: [2023-10-04 18:16:54 +0000] [84603] [INFO] Booting worker with pid: 84603
Oct 04 18:16:54 devsrv gunicorn[84604]: [2023-10-04 18:16:54 +0000] [84604] [INFO] Booting worker with pid: 84604
```

Figure 8-2. *Our hwapp service status*

If we use our web browser and go to port 5000, we will see our familiar "Hello World!" greeting.

From the status, a system administrator could see the hwapp service is active, is a gunicorn-served application that has not yet been enabled, has three worker threads, is located in the **/home/pjh/hwenv** directory, and is available to all IPs at port 5000.

Using NGINX to Share Your Hello World App

We can now have NGINX manage all requests by adding a file to the /etc/nginx/sites-available directory.

But first, we must consider our current configuration where our gunicorn command produces output bound to port 5000 – it could block, or be blocked by, another service trying to use that port. The way around this kind of conflict is to have nginx and gunicorn communicate with socket files created in the application's directory.

Using (uncommenting) the ExecStart line in Figure 8-1 for binding to a unix socket, and commenting out the port binding one we were using, now produces a different output after restarting the hwapp service and checking its status (see Figure 8-3). Now we see our service is bound to unix:app.sock and is also listening there.

```
$ sudo systemctl status hwapp
● hwapp.service - Gunicorn instance for atads project
     Loaded: loaded (/etc/systemd/system/hwapp.service; disabled; preset: enabled)
     Active: active (running) since Wed 2023-10-04 17:38:33 UTC; 5min ago
   Main PID: 83681 (gunicorn)
      Tasks: 4 (limit: 1097)
     Memory: 65.4M
        CPU: 595ms
     CGroup: /system.slice/hwapp.service
             ├─83681 /home/pjh/hwenv/bin/python3 /home/pjh/hwenv/bin/gunicorn --workers 3 --bind unix:app.sock -m 007 wsgi:app
             ├─83682 /home/pjh/hwenv/bin/python3 /home/pjh/hwenv/bin/gunicorn --workers 3 --bind unix:app.sock -m 007 wsgi:app
             ├─83683 /home/pjh/hwenv/bin/python3 /home/pjh/hwenv/bin/gunicorn --workers 3 --bind unix:app.sock -m 007 wsgi:app
             └─83684 /home/pjh/hwenv/bin/python3 /home/pjh/hwenv/bin/gunicorn --workers 3 --bind unix:app.sock -m 007 wsgi:app

Oct 04 17:38:33 devsrv systemd[1]: Started hwapp.service - Gunicorn instance for atads project.
Oct 04 17:38:33 devsrv gunicorn[83681]: [2023-10-04 17:38:33 +0000] [83681] [INFO] Starting gunicorn 21.2.0
Oct 04 17:38:33 devsrv gunicorn[83681]: [2023-10-04 17:38:33 +0000] [83681] [INFO] Listening at: unix:app.sock (83681)
Oct 04 17:38:33 devsrv gunicorn[83681]: [2023-10-04 17:38:33 +0000] [83681] [INFO] Using worker: sync
Oct 04 17:38:33 devsrv gunicorn[83682]: [2023-10-04 17:38:33 +0000] [83682] [INFO] Booting worker with pid: 83682
Oct 04 17:38:33 devsrv gunicorn[83683]: [2023-10-04 17:38:33 +0000] [83683] [INFO] Booting worker with pid: 83683
Oct 04 17:38:33 devsrv gunicorn[83684]: [2023-10-04 17:38:33 +0000] [83684] [INFO] Booting worker with pid: 83684
```

Figure 8-3. *Checking the status of our hwapp. We have multiple workers active, and the service is using a unix socket to communicate with nginx*

If we visit the server's web page at http://A.B.C.D, we will not find the "Hello World!" output, because we have to update nginx to handle the unix socket traffic. This is easily done in a small number of steps:

First, as root, go to /etc/nginx/sites-available.

Second, create a file for your app called **hwapp** containing the block shown in Figure 8-4.

```
server {
    listen 80;
    server_name 146.190.48.240;

    location / {
      include proxy_params;
      proxy_pass http://unix:/home/pjh/hwenv/app.sock;
      }

}
```

Figure 8-4. *A basic NGINX server block that passes requests to http:// A.B.C.D to our hwapp service socket*

This tells **nginx** to send traffic coming into the "/" directory over to our **unix** socket.

Third, create a link to the **hwapp** file in the sites-enabled directory to make it available:

ln -s /et/nginx/sites-available/hwapp /etc/nginx/
sites-enabled

Fourth, restart nginx.

Your hwapp should now show your "Hello World!" message.

(If you encounter issues, check the user and group in the **hwapp. service** file have rw permissions for your app.sock file, and make sure any directory in the path is not blocking access.

We need to take one important additional step. While we're happy our project can be accessed using http://A.B.C.D, this URL should really be reserved for the server's main web interface, which means we should have individual projects accessed elsewhere. A reasonable choice in our case might be from http://A.B.C.D/hello with obvious application to other projects.

To support this style of URL for our project, we need to make two changes to our hwapp.service server block file as shown in Figure 8-5.

```
server {
    listen 80;
    server_name 146.190.48.240;

  location  /hello {
    include proxy_params;
    proxy_pass http://unix:/home/pjh/hwenv/app.sock:/;
    }
}
```

Figure 8-5. The modified **hwapp.service** server block used to support a URL of the form http://A.B.C.D/hello. See the text for important details

There are two critical changes present. The first is the obvious modification of the location block where we use /**hello** instead of /. By itself, this change will not suffice even though it appears intuitively obvious that it should. If you restart nginx, you will likely get a 404 "file not found" error, because "location" works in mysterious ways. To fix this, a second modification to the server block is necessary – note the **:/** added after **app. sock** in the **proxy_pass** argument.

With both changes, restart **nginx** using **systemctl**, and now you should find your "Hello World!" displayed at http://A.B.C.D/hello – success!

Adding the Dashboard Project to Your Server

What about our dashboard project? Based on our preceding experiments, to add our dashboard to the UNIX server, we simply do the following:

1. Build a virtual environment (atadsenv).

2. Install any needed packages.

3. Upload our atads.py, atads_figures.py, and atads_layout.py files.

4. Upload our APT_CSV directory.

5. Upload the folder containing your **style,css** file.

6. Create a **wsgi.py** entry point file.

7. Modify **atads.py** to support **flask**.

Our dashboard's atads.py has an important difference with our hello. py application – in atads.py, we currently have an app as an instance of DASH, but in hello.py it is an instance of Flask. To allow our atads.py version to work with Flask, we need to modify the app instance as shown in Figure 8-6. The basic idea here is we create an instance of Flask called "server," and this is used by the Dash app instance.

```
from flask import Flask
server = Flask(__name__)

app = dash.Dash(__name__,server=server)

...

if __name__ == "__main__":
    app.run(debug=True,host='0.0.0.0')
```

Figure 8-6. *In atads.py, the app instance is modified to work with Flask*

So, with our DASH application, we will use **server** instead of **app**, and in our **wsgi.py** file, we will change the first line to import server instead of app and do **server.run()** instead of **app.run()** – see Figure 8-7.

```
$ cat wsgi.py
from atads import server

if __name__ == "__main__":
    server.run(host='0.0.0.0')
```

Figure 8-7. *The wsgi.py file used for our dash application uses server, not app, since the meaning of app is different; in atads.py, server was the Flask instance, and app was the Dash instance.*

To run our software using Flask, do

flask --app atads run --host=0.0.0.0 --port 8050

and we will see our dashboard at port 8050.

To use gunicorn, do

gunicorn - -bind 0.0.0.0:5000 atads:server

or

gunicorn - -bind 0.0.0.0:5000 wsgi:server

and we will, again, see our dashboard at port 8050.

Creating the Dashboard System Service and Deploying with NGINX

Now that our dashboard is working and can be run using gunicorn, all that's left is to create an atads.service and the necessary NGINX server block. Unlike our hello.py app, where we could either use a port or a Unix socket for nginx to use, we will not use a Unix socket, because Dash is a little more fussy when it comes to sockets.

Figure 8-8 shows the **atads.service** file we will add to **/etc/systemd/system**.

```
$ cat /etc/systemd/system/atads.service
[Unit]

Description=Gunicorn instance for atads project
After=network.target

[Service]
User=pjh
Group=www-data

WorkingDirectory=/home/pjh/atadsenv
Environment="PATH=/home/pjh/atadsenv/bin"

ExecStart=/home/pjh/atadsenv/bin/gunicorn --workers 3 --bind 0.0.0.0:5000 -m 007 atads:server

[Install]
WantedBy=multi-user.target
```

Figure 8-8. *The **atads.service** file*

The service is started with

sudo systemctl start atads.service

For our server block, we will create the file /etc/nginx/sites-available/
atads as shown in Figure 8-9. Since I like to use both our services, I have
merged the hello service into the atads server block.

```
$ cat /etc/nginx/sites-available/atads
server {
    listen 80;
    server_name 146.190.48.240;

  location / {
    include proxy_params;
    proxy_pass http://0.0.0.0:5000/;
    }

  location /atads {
    include proxy_params;
    proxy_pass http://0.0.0.0:5000/;
    }

  location  /hello {
    include proxy_params;
    proxy_pass http://unix:/home/pjh/hwenv/app.sock:/;
    }
}
```

Figure 8-9. *The final server block that allows **nginx** to serve both our
services, **hello.py** using a Unix socket and **atads** using port 5000*

To use this block, delete the unneeded hwapp and link in our new one:

cd /etcnginxsites-enabled
sudo rm hwapp
ln -s /etc/nginx/sites-available/atads .

and start/restart our services:

sudo systemctl start atads.service
sudo systemctl restart nginx

With this design, going to http://A.B.C.D results in seeing port 5000
output directly, while going to http://A.B.C.D/atads and http://A.B.C.D/
hello shows our dashboard and the hello app, respectively. We can now
offer different projects (services) to end users using URLs ending in a
path – a somewhat more elegant solution than using port numbers.

Securing Your Server

To support this book's effort, I created a small inexpensive droplet server, so I could start with a fresh install and build in each part as needed. From a teaching perspective, this mimics the process many readers will follow and avoids confusion where things work too easily because solutions have already been installed on an existing server. Also, using a small server as a sandbox means you have more control and can build, break, and delete things as often as you like.

With your server now saying "Hello World!" or, better yet, serving your project, there will be a sense of achievement in having this resource to support your project and your colleagues' efforts. You will probably be curious about your server traffic and will soon be appalled at how quickly your server will come under attack – the brute-force attempts at logging in through SSH and the requests for web pages and scripts known to have security flaws. There will be thousands of such attempts each day.

At first, you might think it's simply a matter of blocking IPs. This won't work. There are simply too many. Your small server would be more of an IP filter than anything else and waste CPU on this fruitless activity.

A better option will be more expensive – find a hosting service that will block IPs from threatening sites. It's not easy to block the IPs from threatening sites since there are scores of IP subnets assigned to them, and there is no easy universal mask easily implemented. What to do? You do have some options. You should upgrade your server to use HTTPS instead of HTTP. This is easily done using a service like letsencrypt, which you can install for free – although if using it for non-hobby purposes, you should consider donating to them. The install is quite simple, and I won't cover it here since my server is a sandbox without a domain name, but when you use letsencrypt, it will modify your nginx server blocks in sites-available, and if you let it – you should – it will rewrite any incoming HTTP request as HTTPS. This won't stop the volume of attacks, but it will make them less likely to be successful.

Next, make sure your firewall only permits access to the least number of ports possible: SSH, HTTPS, DNS, and FTP mainly.

Your next consideration is easy; is your firewall behind an institutional firewall such as for a college campus or a corporation? If so, you already have a significant level of protection and might simply configure your server to only accept institutional IPs.

Unless you wish to devote a lot of your energy to the server, avoid the temptation to include unnecessary services such as email. If you'd like to receive an email to your server's domain, just use a forwarder at your ISP. It makes things much simpler.

And then there's software like Fail2ban, which is one of the most useful tools to have on your system. It's beyond the scope of this work to give a detailed explanation on how to install it, but the basic idea is it is a utility that can monitor your log files for events like failed SSH login attempts or requests for web files known to have vulnerabilities. Fail2ban is highly configurable, and you could, as an example, specify that repeated failed logins ("the fail") will result in that remote IP being blocked (the ban) for some time. The ban time could be for hours or days. Obviously, the longer the ban, the more IPs there will be to test against for each new connection. Fail2ban is not perfect, and if you're not careful, you could lock yourself out of your system, but it does give a sense of slowing down the assault rate.

Finally, don't forget the most fundamental aspect of server security – being able to recover in the event of an attack or the loss of software or data files. It is very comforting to run a server on a large server hosting site managed by teams of experts, who make sure their servers are properly backed up. Your server is basically a collection of files on another server. So, make a periodic backup or image of your working server. At a moment's notice, you could shut down your misbehaving or corrupted server and start up its image within minutes and be back online. A benefit of this technology is you could test out a major software upgrade using a copy of the server, and if it works, roll it into service. If there are problems, it is easy to back out – fire up the original until you solve the problems on your spare/development server.

Summary

In this chapter, we moved our atads dashboard into the world of server technology, with operating system support to start it automatically and serve it in a multiuser environment, and we addressed security considerations. The dashboard is now a unix service.

Our atads dashboard, now fully deployed and functional, is very straightforward with minimal interaction between widgets in the sense that its widgets were static. In the next chapter, we will show another dashboard that uses important aviation data from the Bureau of Transportation Statistics (BTS), which will illustrate how to have menu items change dynamically and also a capability to allow a user to download the data (subset) used in a chart.

CHAPTER 9

The BTS T100 Dataset: Interacting Controls and Tables

Our dashboard for exploring the ATADS dataset is quite functional and useful. However, because that dataset is essentially a list of stand-alone data for individual airports, with no relational information between airports, our dashboard's controls were easily implemented. There will be times with other projects when controls will interact such as when one menu's items will depend on another's selection, and we will show a way to achieve this. Also, it could be very useful to an end user if they could see the data used to create a chart, displayed as a table, and be able to download it if they wished. For these reasons, we will show a prototype dashboard for another aviation dataset from the Bureau of Transportation Statistics (BTS) – the T100 Domestic Segment data (T100dm). So, in this chapter, we will show a quick prototype based on our previous one, but we will only address the new design challenges involving dynamic menus and table displays.

© Padraig Houlahan 2024
P. Houlahan, *Prototyping Python Dashboards for Scientists and Engineers*,
https://doi.org/10.1007/979-8-8688-0221-8_9

The BTS T100dm Dataset

The T100dm dataset has many columns describing the passenger, mail, and cargo volumes between different airports in the United States. It is developed from monthly reports submitted by airlines. It is beyond the scope of this book to explore it in depth; we will restrict ourselves to those columns relating to dates, origin and destination airports, carriers, and passenger, mail, and cargo (P/C/M) volume.

The challenge we are facing is to design a dashboard that will provide insight into how traffic flows through an airport ("the hub") and between airports ("the segment" linking the hub with a connecting airport).

We can envision an aviation data analyst or a journalist looking for answers to questions like

- What was the volume of traffic (passenger/cargo/mail) that passed through airport X?

- How much of that cargo (P/C/M) was carried by carrier Y?

- How much traffic was there between airports X and Z?

- What were the trends?

- What was the tabulated data (monthly counts)?

- What do the histograms look like?

Prototyping a T100dm Display

Because not all airports have routes between each other, the list of connecting airports will change as the hub changes. If we select a hub airport X, then we have two immediate choices – either consolidate all data for that airport or select a connecting airport Z to allow segment traffic analysis. To support these choices, we need two menus – one for selecting a hub airport and a second for selecting a connector airport (the segment). The hub airport will be drawn from a menu based on all unique airports in the T100dm data, and once the hub is selected, the connector/segment airport list can be constructed. The hub selection menu will dictate what the segment menu shows – conditional upon the mode, since, for example, in Hub mode, the Connecting Airport menu will be empty.

A similar choice will exist when selecting carriers; for either a hub or segment analysis, there will be a possibility of consolidating traffic for all carriers or to specific carriers or to those operating on a particular segment. So, like the Connecting Airport menu, there is a Carrier menu that will be populated depending on whether a hub and/or a connecting airport is specified and on the mode!

Modifying the menus based on hub or segment mode or by carrier is an important service to offer the end user; a reporter might want to know what connections exist between X and Z. It is one thing to know a set of airlines use airport X, but which ones operate between on the X-Z route?

Managing these options, and presenting them meaningfully, is really the heart of our task. We're basically treating the problem as consisting of the display of various modes, with menus adjusted dynamically.

Figure 9-1 tabulates the modes we wish to manage.

Mode	Label	Purpose	Hub Menu	Segment Menu	Carrier Menu
Hub	H	Consolidate all traffic for all carriers	All Airports	-	-
Segment	S	Show consolidated carrier traffic between two airports	All Airports	Connecting Airports	-
Segment by Carrier	SC	Show segment traffic by carrier	All Airports	Connecting Airports	Segment Carriers
Hub Carrier	HC	Show hub traffic by carrier	All Airports	-	Hub Carriers
Carrier	C	Show consolidated carrier traffic for all airports	-	-	All Carriers

Figure 9-1. *The different dashboard modes determine how the menus should be populated*

To implement our design, a mode is specified based on user input, and then menus are populated accordingly, where, for example, with SC, only carriers specific to the hub segment are used for the carrier menu, so for the same hub, if the connecting airport (the segment) is changed, so must the carrier menu. Yes, there's a fussiness in all this, but that's the whole point of the dashboard – to relieve the end user of having to figure this out for themselves from raw data.

Our prototype is shown in Figure 9-2. It is basically a replica of our ATADS dashboard with a modified label and variable name changes with some additional features needed to support this dataset. In what follows, I will mainly address aspects pertinent to this chapter and so will not address all details and functions which can be seen on the online copy.

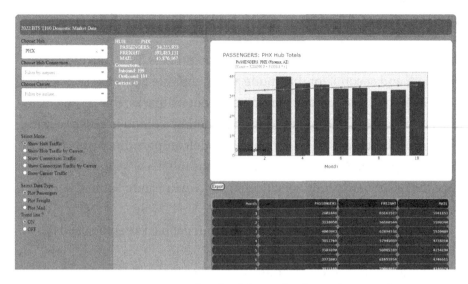

Figure 9-2. *Prototype T100dm dashboard. Changing mode determines the menu items displayed in each of the pull-down menus*

The design is very simple; there are three drop-down menus – for Hub, Connecting Airport (segment), and carrier. There are radio buttons to select the mode, the traffic type, and radio buttons to add a linear equation if desired. A histogram shows the results, and we have also added a table to show the chart's data. Because the table is created using DASH, there is an Export button which allows the user to download the data for use in their own Excel spreadsheet.

So, there are three aspects to our prototype of immediate interest: controlling the menu items, managing the modes, and displaying the table – everything else (stylesheet grids, figures, borders, buttons, etc.) is essentially what we have already covered in previous chapters.

Managing Modes and Interacting Menus

Our menus are populated using a three-step process. First, the mode is specified, and second, for that mode, the menus are activated as needed. Specifically, we pass a mode parameter to our callbacks; this is then used to set intermediate parameters tracking whether a menu is active or not. Third, if a menu is active, it is populated or left empty otherwise. This three-step process is a little complex and was much easier to implement once the dashboard mode specification shown in Figure 9-1 had been completed.

Note that if the mode is not changed, our callbacks can be used to navigate menu items as desired.

We rely on some custom utility functions to extract data of interest from our raw data: **get_hub_df()** when in Hub mode and **get_segment_df()** when in Segment mode.

There are three utilities for extracting carrier data because we can view carrier data for all airports, for a hub, or for a segment (**get_carrier_df()**, **get_hub_carrier_df()**, and **get_segment_carrier_df()**, respectively).

Figure 9-3 shows the mode management code used in the callback routine to process mode changes by flagging whether particular menus are active, extracting the needed data from our raw data, and building appropriate airport, segment, and carrier lists. In Hub mode, for example, the segment and carrier on/off menu flags (**smenu** and **cmenu**) are set to False, the hub's on/off flag **hmenu** parameter is to True, and we only build a dataframe based on the bub **bts.get_hub_df()**.

```
if mode == 'H':
    bts.mode = 'H'
    hmenu = True
    smenu = False
    cmenu = False
    bts.hub = hub
    bts.get_hub_df()
    bts.get_totals_by_month(bts.df_hub)
elif mode == 'S':
    bts.mode = 'S'
    hmenu = True
    smenu = True
    cmenu = False
    bts.segment = connector
    bts.get_segment_df()
    bts.get_totals_by_month(bts.df_segment)
elif mode == 'SC':
    bts.mode = 'SC'
    hmenu = True
    smenu = True
    cmenu = True
    bts.segment = connector
    bts.carrier = carrier
    bts.get_segment_carrier_df()
    bts.build_segment_carrier_list()
    bts.get_totals_by_month(bts.df_segment_carrier)
elif mode == 'HC':
    bts.mode = 'HC'
    hmenu = True
    smenu = False
    cmenu = True
    bts.hub = hub
    bts.get_hub_df()
    bts.carrier = carrier
    bts.get_hub_carrier_df()
    bts.build_hub_carrier_list()
    bts.get_totals_by_month(bts.df_hub_carrier)
elif mode == 'C':
    bts.mode = 'C'
    hmenu = False
    smenu = False
    cmenu = True
    bts.carrier = carrier
    bts.get_carrier_df()
    bts.build_carrier_list()
    bts.get_totals_by_month(bts.df_carrier)
```

Figure 9-3. *The mode parameter is used to determine whether a menu is active or not by setting parameters like* **cmenu** *(for the Carrier menu) to True or False, to extract data used by that mode* **get_xxxx_df()***, and building needed lists of airports and carriers*

The actual populating of our menu items is done in the next step as shown in Figure 9-4. All needed lists of airports and carriers have already been built in the previous step, and they are now used or not – remember inactive menus are left blank.

```python
if hmenu == True:
    hub_options = [{'disabled':False}]
    hub_options =[{'label': i, 'value': i} for i in bts.list_of_airports]
else:
    hub_options = [{'disabled':True}]

if smenu == True:
    segment_options = [{'disabled':False}]
    segment_options =[{'label': i, 'value': i} for i in bts.list_of_segments]
else:
    segment_options = [{'disabled':True}]


if cmenu == True:
    carrier_options = [{'disabled':False}]
    carrier_options =[{'label': i, 'value': i} for i in bts.list_of_carriers]
else:
    carrier_options = [{'disabled':True}]
```

Figure 9-4. If a menu is active, airport, segment, and carrier lists are used to populate menu items; otherwise, the menu is disabled

Figures and Tables

Our plotting methods are almost identical to those used in our earlier dashboard, but because we are using histograms with a trendline option, in case a reader feels this is a substantial change, the relevant methods are shown in Figure 9-5.

```
def add_barchart_trace(self):

        df1 = self.df_selected
        self.fig.add_trace(go.Bar(
        x=df1['Month'],
        y=bts.yvals,
        marker=dict(
            color="blue"
        ),
            showlegend=False
            ))


def do_bar_chart(self,col):

        self.fig = go.Figure()
        self.draw_frame()
        self.get_yvals()
        self.add_barchart_trace()

        if self.show_trend == 'ON':
            self.get_poly(1)
            self.get_linear_equation_string(1)
            self.add_eqn_str()
        self.add_watermark()
        self.add_subtitle()
        self.add_title()
```

Figure 9-5. *Drawing the histograms using the **go.Bar()** trace.*
Because only a linear trend is used, our customized polynomial
*solution is set to first order using **get_poly(1)***

The whole point of having modes and menus is to extract a subset of
data from our raw data and in our code. In addition, all modes use the
get_totals_by_month() method to produce monthly totals suitable for
plotting, and it will insert a column with months ranging from 1 to 12. For
example, the **get_totals_by_month(df_hub)** method extracts data from
the hub's dataframe **df_hub** and sets the **df_group_by_month** dataframe.
And similarly for segment and carrier reporting (see Figure 9-6).

147

```
    ...
    def get_totals_by_month(self,df_in):
        res=df_in.groupby('MONTH')[['PASSENGERS','FREIGHT','MAIL']].sum()
        nrows = res.shape[0]
        count = list(range(1,nrows+1))
        res.insert(0,'Month', count)
        self.df_group_by_month = res

    ...

        html.Div(className = 'hub_table',
                    children=[
                    dcc.Graph(id="sgraph"),
                    dash_table.DataTable(id='hub_tbl',
                        style_header={
                        'backgroundColor': 'rgb(30, 30, 30)',
                        'color': 'white'
                        },
                        export_format = "csv",
                        style_data={
                            'backgroundColor': 'rgb(50, 50, 50)',
                            'color': 'white',
                            'border': '1px solid blue'}

                    )
                    ]),
    ...
@app.callback(
    ...
    Output('hub_tbl',            'data'),
        ...

    if mode == 'H':
        ...
        bts.get_totals_by_month(bts.df_hub)
        ...
        data = bts.df_group_by_month.to_dict('records')

        ...

        return ..., data,...
```

Figure 9-6. get_totals_by_month(df_hub) *is used in hub mode to set the working dataframe **df_group_by_month** which is then used to create dictionary **data** and returned by the callbacks to the layout's **DataTable***

Filling the table is then achieved by creating a dictionary called **data** from **df_group_by_month** and returning it from the callback to the table layout.

Please note a word of caution. Python is very popular among data analysts, and it offers some extremely powerful functions for summarizing dataframe contents by aggregating or splitting or grouping entries, and these capabilities can be very confusing to work with. Unless you are a specialist working regularly with these tools, and hence intimately familiar with them, I would suggest using the simplest and most brute-force solutions you can get away with since they will likely be the easiest to work with at a later point when you need to refresh your understanding of how they functioned.

Summary

Complex dashboards will require complicated interactions among their widgets, such as needing dynamic menus. In this chapter, we showed one way to organize how menus are populated depending on what functionality (mode of operation) we were pursuing. We also demonstrated how to make the data used to produce a chart available to the user, who might need to do their own analysis using Excel.

In the next chapter, we will show how to encapsulate and share your project through building a web portal that offers users not just access to the dashboard but also to documentation and to forums where they can learn from each other and contribute to the knowledge base.

CHAPTER 10

Creating a Web Portal

Having created a dashboard that resides on your server, you might want to consider managing how it is presented to your colleagues and end users by creating a web portal, by which I mean a web presence which serves as the entry point to your projects and also as a location where you can provide a context for your projects that could include links to blogs, documentation, and other project software.

Your web portal could be as simple as a basic HTML page with a welcome header and links to your project's dashboards and other documentation. It might include corporate or college logos and maybe some graphics to give it a little finesse. By keeping it simple, your life will be easy.

In Figure 10-1, I show a primitive but functional portal to share dashboards with students. It's not very fancy and was built using very rudimentary HTML, but it gets the job done. It has the advantage of being easy to implement, and enhancements can be deferred to a later time.

© Padraig Houlahan 2024
P. Houlahan, *Prototyping Python Dashboards for Scientists and Engineers*,
https://doi.org/10.1007/979-8-8688-0221-8_10

AVOPSinsight Web Portal

(A resource for airport and airline professionals)

Documentation and Blog

Airport Operations: FAA Air Traffic Activity System (ATADS).

FAA ATADS Dashboard

NEW!!!!! Just added June 2023!!!!

Bureau of Transportation Statistics - T100 Domestic Market dataset.

BTS T100 Domestic Market Dashboard

Figure 10-1. *A primitive but functional welcome page for end users, giving them options to go to different project areas*

In case you are new to HTML coding, and just want to get going, the code used to build the portal is shown in Figure 10-2.

```
<!DOCTYPE html>
<html>
<body>

<h1>AVOPSinsight Web Portal</h1>
<h3>(A resource for airport and airline professionals)</h3>


<a href="https://avopsinsight.com/wordpress">Documentation and Blog </a>

<br>
<h3>Airport Operations:  FAA Air Traffic Activity System (ATADS).  </h3>
<a href="https://avopsinsight.com:8080">FAA ATADS Dashboard </a>

<br>
<br>
<br>
<h4> NEW!!!!! Just added June 2023!!!! </h4>
<h3>Bureau of Transportation Statistics - T100 Domestic Market dataset.  </h3>
<a href="https://avopsinsight.com:9090">BTS T100 Domestic Market Dashboard </a>
</body>
</html>
```

Figure 10-2. *The very basic HTML code used to generate the page shown in Figure 10-1*

There is one notable advantage of using such a simple code – there's not much that can go wrong in the sense all browsers will render the page. And, this page's directory, being at the top of your project web tree, would be a good place to implement login and access controls (perhaps using a .htaccess file) which would protect your whole project tree.

But a web portal can be so much more, as shown in Figure 10-3 where I used WordPress to build out the portal. My considerations in designing the portal were

1. I wanted to have as uncluttered a look as possible.

2. I wanted essential features, such as links to dashboards, immediately accessible to the visitor.

3. Visitors should be enticed to explore dashboards by seeing some of their graphical charts.

4. There should be an obvious location for documentation describing information about the dashboards and how to use them.

5. For users, a registration and blog area for discussions – mainly with the intent to support student courses.

6. No ads. Yes, that is a heretical position to hold, but I can live with it for now.

As of late 2023, this web portal is still under construction, but it is functional and extensible.

Figure 10-3. *The AVOPSinsight.com web portal is intended to provide an attractive first impression for students and end users, where they can quickly find dashboards of interest and access other resources*

Troubleshooting WordPress

WordPress is an extremely powerful social media application, and being powerful, it can be daunting trying to achieve a finessed result. There are detailed online descriptions on how to first install MySQL, a necessary prerequisite, so I won't address them here.

When installing WordPress, you will have to decide what theme to use, where themes define the WordPress look and feel, such as how many menu bars or vertical columns are presented. There is a whole industry of companies offering an enormous number of choices; themes and fads come and go, and you could spend vast amounts of time trying to find a nonexistent perfect theme. I would recommend picking a basic one offered by the WordPress designers until you feel like you need a more complex one and you're ready to take it on as a major project.

Once you have selected a theme, it can be a real nuisance to find your site looks quite good, but an annoying quirk persists. While many themes are free, I have found it next to impossible to find a perfect one – there always seems to be some maddeningly frustrating deficiency – which means you either have to live with it, learn how to fix it, or pay the designers to customize it. Being a cynic, I suspect the deficiencies are intentional to encourage business!

So how do you fix one of these quirks? There are a bewildering amount of code and stylesheet/CSS parameters underlying most themes; your challenge is to identify the settings specific to the quirk you are trying to fix, and WordPress allows you to add parameters which will override the theme. I will now demonstrate how I tracked down the setting behind one particular quirk using Chrome's Developer Tools and adjusted it to get a more pleasing result. The process is similar to how we figured out which variable names to use when we explored screen-scraping ATADS data in an earlier chapter, but WordPress themes can be very complex, so we will undertake a more in-depth description of how to work with the Developer Tools.

155

For my portal, I used the **twentysixteen** theme, and while I liked its clean layout style, it had a very ugly aspect in that the listed items along the right-hand sidebar were too widely spaced – see Figure 10-4.

To fix this line spacing issue, we can use Google Chrome's toolbox to identify the source of the "problem," and then since WordPress allows users to enter custom CSS commands, we try custom CSS commands to attempt a fix by overwriting default parameters.

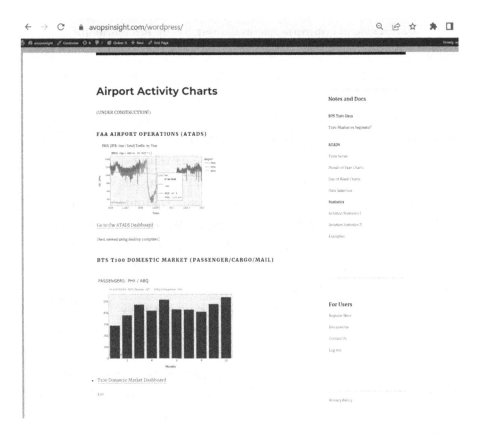

Figure 10-4. *The default layout for the twentysixteen theme yields an unsatisfactory excessive spacing between list items along the right side*

In Chrome, go to the top-right corner, click the three vertical dots, and locate the Developer Tools area as shown in Figure 10-5. You might now be presented with a screen that is overwhelming such as that in Figure 10-6.

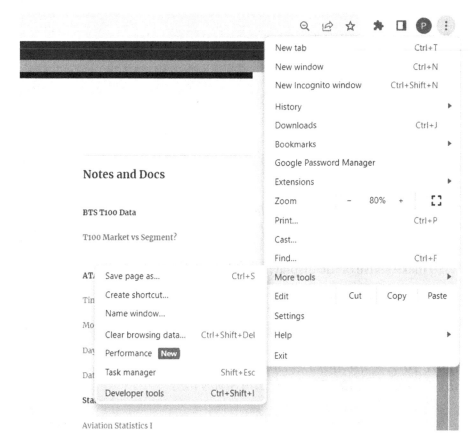

Figure 10-5. *Finding Chrome's Developer Tools*

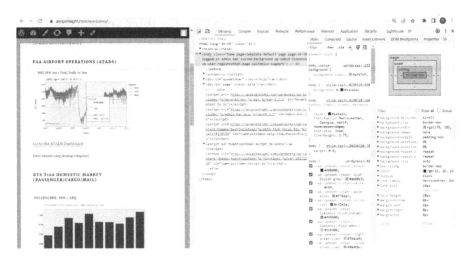

Figure 10-6. *Developer Tools reveals the thousands of parameters defining your theme's look and feel. To make room for its output, the web page is squeezed into the left column*

To make sense of the display, note the two horizontal menu/tab bars along the top (see Figure 10-7).

Figure 10-7. *Chrome's Developer Tools has a complex selection of tabs. We focus on the Elements and Styles tabs*

The top one shows we are working with elements, and the second shows we have selected the CSS Styles view. If you move your cursor up and down the second (Elements) column, your web display background will show a blue background as you highlight various elements. You will have to click the small black triangles to open up hidden layers. Keep doing this until you have identified the area of interest.

Figure 10-8. *Moving your mouse cursor highlights blocks in the left column display. If the blocks are too large, click the small triangles to go deeper into the styles*

After some testing, I found elements associated with the lists (see Figure 10-8), and this seemed like a good place to examine. Clicking an element causes three small dots to appear which mark which element is selected, and notice on the far right, I see an intriguing padding statement (Figure 10-9). If I place the cursor over this statement, a small circle with an arrow appears, and if I click that, then at the top of the styles column I see an associated widget style (Figure 10-10).

Figure 10-9. *On the far right, we see a padding specification associated with our selected styles*

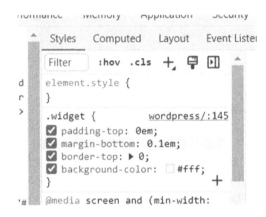

Figure 10-10. *We have now identified a widget associated with our list item*

Notice some blue checkboxes appear if you place your cursor near the widget entries, and you can use these checkboxes to toggle features on and off. In this case, I can toggle the padding-top on or off and see the effect on the leftmost column showing my blog lists, and indeed the list entry spacing changes as I toggle the padding. (In this case, the toggling was activating/deactivating the padding-top:0em I had previously set to fix the spacing.)

So, we have successfully identified the elements causing the unsightly spacing. To fix the problem, we now add an override into our WordPress customization area. Go to the WordPress dashboard, and find the Additional CSS area for your theme, and force the widget's padding into the form you want (see Figure 10-11). In my case, setting the padding-top to 0.0em worked! From this point on, the blog will show nicer spacing that better defines my lists visually.

Figure 10-11. *The entries placed in the WordPress Additional CSS field that overrides the default line spacing and creates a more pleasing list grouping*

Summary

In this chapter, we explored wrapping our project in a web portal which allows us to manage and centralize our efforts. We can offer visitors documentation and blog forums for them to learn from each other and to contribute to our project's knowledge base – a powerful, self-contained encapsulation of our work and perhaps also of our team's. We used WordPress because it is well placed to do just this; however, WordPress can be very nuanced, and so we demonstrated how Google's toolbox can be used to track down troublesome design or layout quirks, an essential skill for anyone relying on WordPress. The payoff of course is a web portal that can be visually attractive and very functional, a valuable asset for colleagues and team members.

CHAPTER 11

Using Our Dashboard for Data Visualization and Analysis

Now that we have a working dashboard, let's address the issue of how best to use it and see how it performs against the ATADS dataset. You will likely have your own dataset, but there will be strong similarities in how data – especially time series – should be explored. While I write this chapter with aviation business students, researchers, and professionals in mind, the non-aviation specialist can see how useful our dashboard's tools are, see how we use features such as spectra and trends, and perhaps also think about how our strategies/interpretations could be similarly applied to their project data.

Understand your data. Don't just treat it as a mere collection of measurements. Spend some time thinking about what was measured and why. For example, in the ATADS data, there are categories for local and itinerant traffic. For itinerant, we would naturally think in terms of air carrier and air taxi, but it also includes small aircraft operations, all between airports. What then is interesting about local? There are many operations where aircraft return to their point of departure – EMS helicopters, media (weather and traffic), pilot training, and firefighting. A remote airport might be fairly quiet for most of the year, but there could be event-driven activity such as fires and airshows!

© Padraig Houlahan 2024
P. Houlahan, *Prototyping Python Dashboards for Scientists and Engineers*,
https://doi.org/10.1007/979-8-8688-0221-8_11

We will demonstrate how we can use our dashboard to explore different attributes, from using spectra to detect periodic patterns to comparisons between major hubs based on where they are located, exploring different airport types, and disaster studies. The cases explored here are selected to demonstrate ways of approaching, documenting, characterizing, and presenting data in a manner suitable for reports and web content.

Airport Type, Trends, and Location

An airport's location can influence its working. As an example, along the East Coast of the United States, airports are more likely to be impacted by hurricanes, unlike the West Coast. Figure 11-1 shows the air carrier traffic for three major hubs, suggesting they are operated close to capacity since the graphs tend to be fairly constant. End-of-year holiday slowdowns for Christmas and Thanksgiving days are evident. Los Angeles (LAX), being on the West Coast, experiences few major interruptions, unlike Colorado's Denver (DEN) and New York's John F. Kennedy (JFK). With charts like this, one could document storm frequency and impact for (disaster) planning purposes.

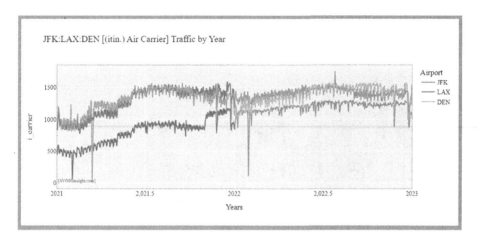

Figure 11-1. *Three major hubs in the United States show their recovery from the pandemic. The graphs suggest the airports are close to being saturated. Some show significant short-term drops not shared by others, indicating closures/slowdowns caused by storms and hurricanes*

While airports like JFK seem congested, Arizona's Grand Canyon (GCN) airport, relying heavily on local tour flights – especially using helicopters – we would expect to be busiest in the summer months. However, looking at six recent years of operations suggests its business changed after the COVID-19 pandemic since the pre-pandemic activity was significantly greater (see Figure 11-2).

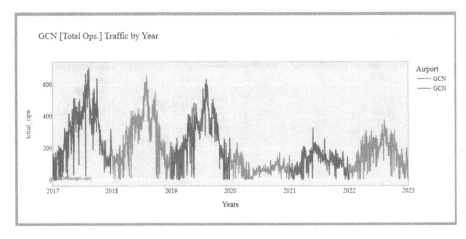

Figure 11-2. *Grand Canyon airport (GCN) mainly caters to offering tourists air tours. While activity has been building since the pandemic, it is substantially below pre-pandemic levels*

Airshows and Seasonal – Using Spectra

(Because spectra are normally used in scientific and engineering studies, it is novel to apply them to business activities, and it is worth discussing their uses to better explain them. The insights they provide are more like curiosities, and the real purpose they serve is to help the aviation professional or researcher better appreciate their data. Since spectra separate various periodicities in a dataset, we initially consider one of the simplest – a repeating pulse in the data – such as what occurs with annual airshows.)

With airshows, we expect a surge in civilian aircraft, and perhaps military, traffic, such as the famous annual one at Oshkosh (OSH). Figure 11-3 shows the VFR traffic that visits this airport during its annual fly-in. Most small aircraft pilots will arrive under VFR conditions since IFR is more demanding and dangerous for those not well practiced. Also, there is such a high density of traffic in the area that trying to handle so many

IFR approaches would simply slow things down unacceptably. The figure shows us the fly-in was canceled during the pandemic but returned to normal thereafter. Peak operations during the busiest week reached 2500 operations in a day. That's almost one operation every 30 seconds.

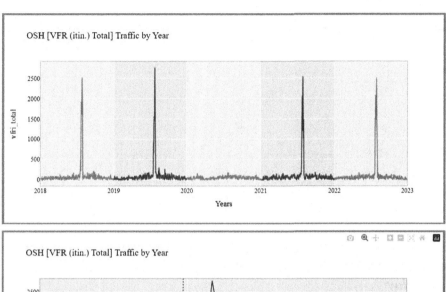

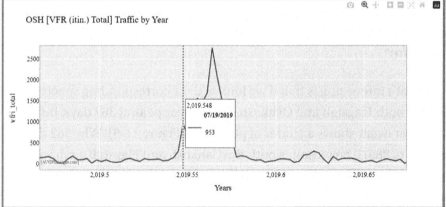

Figure 11-3. *The top panel shows a multiyear view of the annual Oshkosh fly-in traffic. The pandemic caused its cancelation, but it returned successfully afterward. In the bottom panel, we used the cursor to select the peak in 2019, and we can see how there were between 1000 and 2500 operations in the course of a week*

A smaller airshow is held at my hometown airport, Flagstaff (FLG). This is smaller and is more of a show than a fly-in. It makes an interesting comparison with Oshkosh because most of the activity happens on the day of the event (see Figure 11-4).

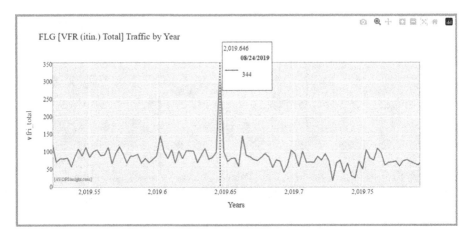

Figure 11-4. *Flagstaff (FLG) has an annual airshow which is more of a display than a fly-in. As a result, its peak traffic happens over a single day*

What's interesting is that if we look at their corresponding spectra, we find both Flagstaff and Oshkosh have a clear peak at 362 days, but the larger event shows a clutter of peaks (see Figure 11-5)! Why 362 days instead of 365? The reason is both the Oshkosh and Flagstaff airshow spectra are based on a multiyear data selection, and there is no guarantee of a precise interval of 365 days between annual events, since organizers are probably trying to pick a weekend/day-of-week instead of an exact 365-day interval.

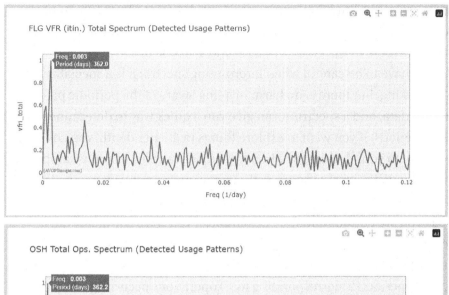

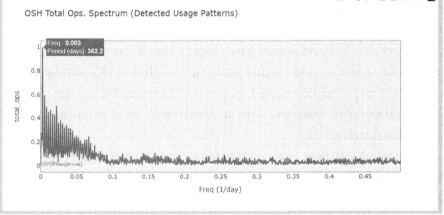

Figure 11-5. *Since Flagstaff's airshow is largely a one-day event, its spectrum (top) shows a strong peak with a period of about a year (362 days). The Oshkosh (OSH) fly-in is a weeklong annual event, and its spectrum (bottom) also shows a 362-day peak and many smaller ones*

Why are there so many peaks in the OSH spectrum? I suspect the algorithm is detecting periodic intervals where the main fly-in is being compared to seasonal and recreational traffic, that is, there is significant seasonal traffic, but it perhaps appears overwhelmed by the fly-in's spike. And, for example, traffic arriving a week before and leaving a week later will contribute to a period of 14 days and also one of 350 days.

We have to be careful when interpreting spectra; it is a specialized undertaking, but there is no harm in being aware of the periodic patterns in your data, and a spectrum can give you a quick way to determine the main periods. If you wish to explore things in greater depth, you could add a feature to your dashboard where you inject patterns of known amplitude and periodicity – just to see how they show up in your spectra; the idea here is to add a signal you understand (because you created it) near the peak you are curious about. For example, if you were curious about a seven-day period, adding a series to your dataframe consisting mainly of zeros, but with every 12th entry being 10, for example, should produce a spike at period 12, corresponding to 10 operations occurring every twelfth day. If your nearby injected signal peak with amplitude 10 is similar to the seven-day one you are interested in, you could reasonably say the seven-day period signal is consistent with an activity pattern of amplitude 10 operations per day.

For a regular airport, one could look at Phoenix Sky Harbor (PHX)'s Total Operations and visually see there are short-term (weekly) cycles and seasonal ones. The spectrum is quite striking, and we see there are distinct 7-day, 3.5-day, and 2.3-day periodic operations corresponding to weekly, twice weekly, and three times a week operations (see Figure 11-6).

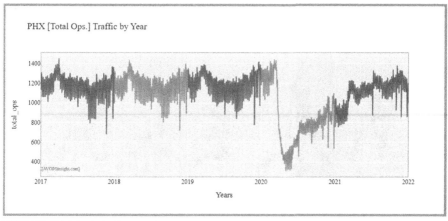

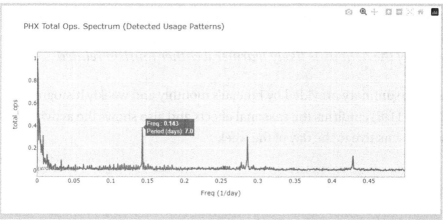

Figure 11-6. *Phoenix (PHX)'s spectrum (bottom) shows strong 7-, 3.5-, and 2.3-day cycles corresponding to 1, 2, and 3 times per week. Remember, two times the frequency, halves the period, etc. The Total Ops. Chart (top) shows sr=tronf weekly variations, seasonal effects, and the pandemic's influence*

While Phoenix shows a seasonal effect, if we consider more northerly airports exposed to harsh winters, we see a very different annual activity pattern such as that for Juneau (JNU), Alaska (see Figure 11-7). Almost certainly, the activity is tourism related.

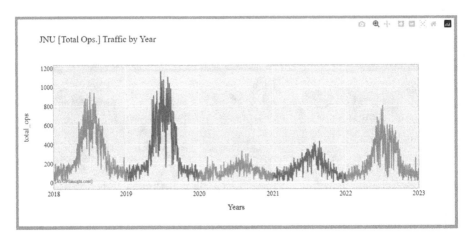

Figure 11-7. *Five years of Total Operations for Juneau (JNU). There is clearly a strong seasonal effect still present during the pandemic. Much of the activity is likely warmer weather tourism related*

The summary provided by Juneau's monthly and weekly histograms (Figure 11-8) confirms the seasonal effects and also shows the activity is weakly sensitive to the day of the week.

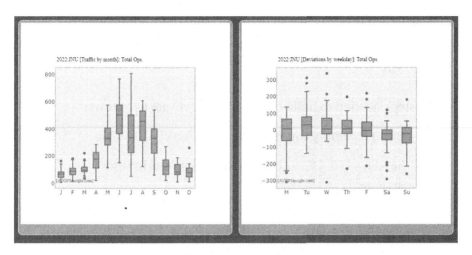

Figure 11-8. *The monthly chart shows a cleaner representation of Juneau's Total Operations, emphasizing the seasonality, while the weekday chart suggests a continuous daily routine, largely independent of the day of the week*

Incorporating Models

Sometimes, when exploring a dataset, it can be very useful if we can create a model. This does two things: it gives us opportunities to test datasets against theories/models, and a good model will usually present a much simpler summary of the observations. Since we are mainly concerned with time series data, our models will need to be constructed from Python lists.

To take a concrete example, let's try and understand Oshkosh's spectrum (Figure 11-4). Our goal here is to see if we can create a dataset with similar characteristics to its Total Ops. Chart and, by tweaking the model, see if we can account for the spectrum.

Here are the main attributes I can see visually from the Total Ops. Chart (Figure 11-3): there is a large annual burst of activity lasting for a week of amplitude 1000+ ops/day, and there is a seasonal behavior that peaks in the summer.

173

We will treat these as two signals that need to be added together, and our task is to now model these signals separately. To do this, let's build the models using an IDE (integrated development environment) console for testing.

The steps needed to build the fly-in's list (yf) are as follows:

1. Build an empty list of length 365 using yf = [0] × 364

2. Set days 178–182 to 1500 as a crude model for the activity

3. Replicate for the five-year interval: yf = yf × 5

yf is now a 1820-long list of zeros except for the group of elements every 364 days of size 1500.

To build the seasonal background yb, we will assume there is an overall amplitude of 100 that varies by season:

1. Create a week's worth of data: yb = [100] × 7

2. Make a year's worth: yb = yb × 52

3. Add a block of activity for the fly-in days

4. Include an annual sine wave modifier

5. Replicate for 5 years using yb = yb x 5

This model assumes the fly-in impact is about 1500 daily operations for 5 days, and the non-flying traffic is 0 in the winter but peaking at 100 in midsummer. We also assumed a year was 364 (52 × 7) days for convenience. In general, though, we would like to be able to adjust these, so we encoded them as parameters **a1** and **a2** used by a method in our **atads_figures.py** file (see Figure 11-9).

```
# Build a model where for 5 years Oshkosh fly-in traffic adds a2 operations for days 175 - 182
#
# Include a background non-fly-in activity that is a sine function that peaks at level a1
#
# assume years are exactly 52 weeks long (364 days)

import math                                  # using the math libs

    def build_oshkosh_model(self,a1, a2):
        yf = [0]*364                         # build a year of zeros
        for i in range(175,182):             # set daily ops for a week to a2
            yf[i] = a2

        yf = yf*5                            # replicate to a 5 year span

        yb = [a1]*364                        # build a year of data background data


        for i in range(0,364):               # add a sinusoid modulation
            yb[i] = yb[i]*math.sin(3.14*i/364)

        yb = yb*5                            # replicate to a 5 year span

        y_osh = []                           # initialize our model output list
        for i in range(1820):                # combine the fly-in and background ops.
            y_osh.append(yf[i] + yb[i])

        return y_osh

    def update_spectrum(self,airport_list, yr_list, active_variable):
        ...
        for i in airport_list:

                ...
                y_vals = self.build_oshkosh_model(100,1500)    # Testing a1=100 and a2=1500 combo
                ...
```

Figure 11-9. *The code used to implement our model where the fly-in adds **a2** ops per day, and the non-fly-in activity is a sinusoid peaking at **a1** ops midyear. In the update_spectrum() method, I manually set a1 and a2 to 100 and 1500, respectively*

With this module, we can change the a1 and a2 parameters to see the effect on the model spectrum.

The results are very interesting and can be seen in Figure 11-10 where I show models where the fly-in was comparable to the background traffic (**a2** = 150) and also where it was substantially greater (**a2** = 1500).

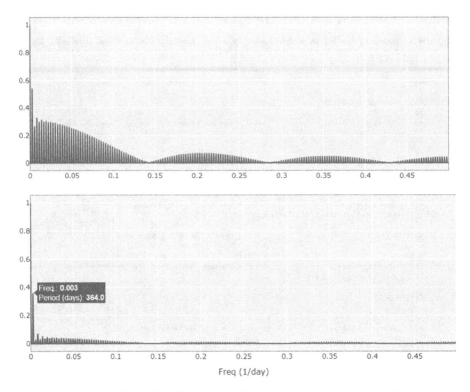

Figure 11-10. *When the fly-in amplitude was 15 times the local background (non-fly-in), the complicated set of peaks emerges in the spectrum (top). With a model where it was only 50% greater, the spectrum's main annual peak dominates (bottom)*

There are many variations we could try to our model – add an offset to the background so there is nonzero traffic in the winter, and try something other than a flat week of activity for the fly-in. We could try a short period signal for regular airports to model weekly activity and so on.

The lesson here is that Python has some very elegant ways to create lists of data that allow you to construct models for the observations, and they can be used to develop insight into your datasets – in this case, nicely explaining how fly-in activity compared to non-fly-in influences the airport's spectrum.

Media, Presentations, Reports, and Projects

Having nice customizable graphics helps you understand your data, but being able to use them to support project and company objectives is also a considerable benefit. We live in a media age, where people like to see attractive images, and there is pressure to continuously create new web content, especially for public consumption. Reports and white papers are management tasks that can also benefit from attractive graphics. In academics, researchers and students can incorporate them into journal articles and class projects. Because the ATADS dataset is updated monthly, our dashboard could play a regular role in creating new content and supporting academic and business activities.

In the following text, I present a list of projects suitable for airport staff facing a need for web content creation and report writing and airport management students writing theses and white papers. Obviously, the context (web/corporate/academic) will dictate a task's desired outcomes, but the list does convey how useful a dashboard like ours can be for supporting such projects. (When I suggest documenting or exploring, I envision projects that students could do where they build reports incorporating suitable charts and numbers providing quantitative results.)

To demonstrate the possibilities, I did a Google search for "Alaska wildfire smoke" to see if there was an event that would show up in our ATADS data. I immediately found a NASA page for a May 2002 fire near Fairbanks (FAI), Alaska (see Figure 11-11).

🔒 earthobservatory.nasa.gov/images/9603/fires-and-heavy-smoke-in-alaska

This page contains archived content and is no longer being updated. At the time of ✕
publication, it represented the best available science.

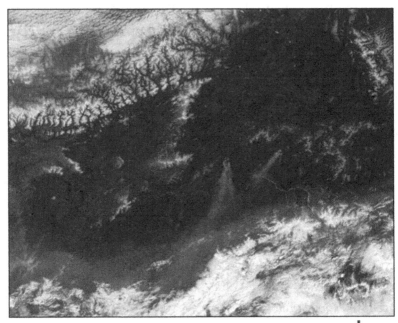

May 27 12:00:00, 2002   ⬇ JPEG

Heavy smoke coming from forest fires around Fairbanks, Alaska, has drifted south and
appears to be lined up with the clouds at the bottom of the image. The fire just right of
center is the MP 78 Elliott Highway Fire, and to its east is the West Fork Chena Fire. Both
of the fires continue to spread, crossing rivers and roads, and threatening structures. This
image was acquired by the Moderate Resolution Imaging Spectroradiometer (MODIS) on
the Terra satellite on May 27, 2002.

Figure 11-11. *A NASA page describing smoke near Fairbanks (FAI)*
in May 2002

Using our dashboard, we can see there was indeed a measurable drop
in April/May 2002 operations (Figure 11-12).

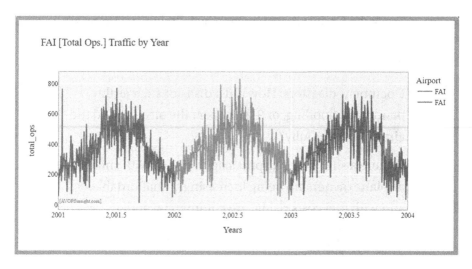

Figure 11-12.   *Three years of data for Fairbanks (FAI), spanning the years 2001–2003. There is a noticeable dip around late April 2002, not present in adjacent years, probably caused by the smoke from wildfires*

Possible projects and activities:

- Document trends for growth, noise, and congestion studies: Do the charts suggest the airport is stagnant or saturated?

- Document the nature of the airport: Is it mainly for small aircraft? Airlines? Mainly for tourism or very seasonal?

- Explore the impact of rain and snowstorms, hurricanes, and fires (smoke): Document the incidences of historical events – these can be identified by noting significant dips in operations' data and doing Google searches for those dates and cities. By what percentage did storms degrade throughput?

- Track the impact of a hurricane by following its impact on airports along its path; the dip in activity will change day by day as the storm moves north.

- Document disasters: How did a disaster such as the pandemic, flooding, or 9-11 impact the airport? Did the airport recover fully?

- Is there a shift in the airport's operations? With small airplane ownership being increasingly unaffordable, and with fuel costs rising, are commercial and noncommercial operators equally affected?

- Estimate the activity levels for the next three years using trend data.

- Compare an airport with another: Identifying peer airports can provide useful insight into trends and performance.

- Develop models for airport operations using Python tools to characterize observed historical trends and explain their spectra.

While many of these questions are of interest to all, the fact that there are more than 500 airports with data that change monthly allows business school faculty to offer a very diverse range of projects, with fresh variations possible each semester.

In my opinion, projects like these can provide useful material for journalists and other non-aviation specialists and airport supporters and are appropriate for corporate websites and social media.

Summary

This chapter showed how we can use our dashboard to explore different attributes, such as using spectra, exploring different airport types, and disaster studies. These cases demonstrated how approaching, documenting, characterizing, and presenting data work for reports and web content.

CHAPTER 12

Afterword

Being able to build dashboards to support your colleagues or students is an extremely empowering capability. With a good design, you can present data and analysis tools to them to make their work as effortless and pleasing as possible. In learning how to design and build dashboards and how to deploy them, you are mastering technologies that allow you to encapsulate and share expertise.

There were times when I naively tried sharing raw code demonstrating some aspect of a course I was teaching and encountered frustration when code that worked fine on one OS didn't work at all on another, where some students had sufficient skills to install software on their laptops and others had no idea on what to do with missing libraries or OS resources. The solution at the time was to install my software on each computer in a student computer lab. With the techniques concerning deployment demonstrated in this book, those kinds of issues are a thing of the past – the students could now simply go to your web portal.

What's missing from this book is what to do if you create a dashboard in wide demand. The easiest answer is to use a technology like Kubernetes (Google also offers this kind of capability) that can load balance and scale your dashboard to serve many users. For small project teams, you are unlikely to need them, but if you are filling an important and high-demand niche, your peers and students would benefit.

© Padraig Houlahan 2024
P. Houlahan, *Prototyping Python Dashboards for Scientists and Engineers*,
https://doi.org/10.1007/979-8-8688-0221-8_12

Don't be afraid to divide and conquer. If you can create one dashboard, you can create others. While it's always a judgment call, it's probably best to separate out different aspects of your team's work into different dashboards, letting your web portal offer your users a menu of solutions, then to blend them all into one major project. And remember, each time you create a stand-alone dashboard, not only will you have the satisfaction of completing a self-contained project that encapsulates your team's capabilities and solutions, but you will also be setting your project up for stability – users don't really like to see the software they have come to rely on change. Creating dashboards can be a rewarding and satisfying outlet for your creative skills that essentially self-document and stabilize your work, allowing you to concentrate on new challenges. They really are the next step beyond simply coding a solution to be run at the command line from a particular directory; once you start creating them, you probably will not settle for anything less in the future.

Finally, the whole point of creating a dashboard is to share data and hopefully gain a greater depth of understanding, which is why I took many opportunities to demonstrate interesting phenomena and techniques and why I think it's a great way to help students learn – and hopefully develop an appreciation for – the art of communicating technical topics supported with compelling graphics and quantitative results.

APPENDIX A

Utilities for Managing ATADS Data

Since the Federal Aviation Administration's ATADS data is refreshed monthly, your ATADS dashboard also needs to have its data periodically updated. I manage the data by keeping downloaded and derived files and the needed utilities in a folder (ATADS_DATA_UTILS).

I use three main subfolders:

> ATADS_XLS: A directory containing downloads of annual Excel files, for example, atads2020.xls

> ATADS_CSV: A directory containing annual files converted from xls to csv, for example, atads2020.csv

> APT_CSV: A directory containing csv data for individual airports, for example, JFK.csv

The three core utilities provided in Figures A-1 through A-3 are

- Figure A-1: **atads_scrape.py** (a utility for downloading data)

P. Houlahan, *Prototyping Python Dashboards for Scientists and Engineers*,
https://doi.org/10.1007/979-8-8688-0221-8

```python
########################################################################
#
# Load libraries
#
########################################################################


from selenium import webdriver
import time
from selenium.webdriver.common.by import By

########################################################################
#
# USER SETTING
#
########################################################################

do_year = "2000"

########################################################################
#
# Set the target page URL
#
########################################################################

urlpage = 'https://aspm.faa.gov/opsnet/sys/airport.asp'
print(urlpage)


########################################################################
#
# Tell Chrome where to save downloaded data
#
########################################################################

options = webdriver.ChromeOptions()
profile = {
    "download.default_directory": "C:/windows/Users/pjhmx/Desktop/",
    "download.prompt_for_download": False
    }
options.add_experimental_option("prefs", profile)
```

Figure A-1.   *Screen-scraping demo code: atads_scrape.py*

```
####################################################################
#
# Load the Chrome driver
#
####################################################################

driver = webdriver.Chrome(options=options)

#driver = webdriver.Chrome()

# get web page

####################################################################
#
# Activate the driver
#
####################################################################

driver.get(urlpage)


####################################################################
#
# Wait for the driver to open and load the target page in a browser
#
####################################################################

time.sleep(30)
```

Figure A-1. (*continued*)

```
#######################################################################
#
# Set BUTTON:REPORT parameters
#
#######################################################################

mylink=driver.find_element(By.ID,"b_repOpt")
print(mylink)
mylink.click()

driver.find_element(By.CSS_SELECTOR,"input[type='radio'][value='msexcel']").click()
driver.find_element(By.NAME,'nosubtot').click()
driver.find_element(By.NAME,'ifr').click()
driver.find_element(By.NAME,'vfr').click()

#######################################################################
#
# Set BUTTON:DATE Parameters
#
#######################################################################

mylink=driver.find_element(By.ID,"b_dSelector")
print(mylink)
mylink.click()

driver.find_element(By.ID,'RangeOption').click()
driver.find_element(By.XPATH,"//select[@name='fm_r']/option[text()='Jan']").click()
driver.find_element(By.XPATH,"//select[@name='fy_r']/option[text()="+do_year+"]").click()
driver.find_element(By.XPATH,"//select[@name='fd_r']/option[text()='1']").click()

driver.find_element(By.XPATH,"//select[@name='tm_r']/option[text()='Dec']").click()
driver.find_element(By.XPATH,"//select[@name='ty_r']/option[text()="+do_year+"]").click()
driver.find_element(By.XPATH,"//select[@name='td_r']/option[text()='31']").click()
```

Figure A-1. (*continued*)

```
##################################################################
#
# Set BUTTON:FACILITIES parameters
#
##################################################################

mylink=driver.find_element(By.ID,"b_locOpt")
print(mylink)
mylink.click()

##################################################################
#
# Set BUTTON:FILTERS parameters
#
##################################################################

mylink=driver.find_element(By.ID,"b_addOpt")
print(mylink)
mylink.click()

##################################################################
#
# Set BUTTON:GROUP parameters
mylink=driver.find_element(By.ID,"b_groupSelector")
print(mylink)
mylink.click()
```

Figure A-1. (*continued*)

```
##########################################################################
#
# Build data request
#
##########################################################################

js = "addFld('DATE')"
driver.execute_script(js)
js = "addFld('LOCID')"
driver.execute_script(js)
js = "addFld('STATE')"
driver.execute_script(js)
js = "addFld('REGION')"
driver.execute_script(js)
js = "addFld('DDSO_SA')"
driver.execute_script(js)
js = "addFld('CLASS_ID')"
driver.execute_script(js)

##########################################################################
#
# Submit data request
#
##########################################################################
mylink=driver.find_element(By.ID,"b_Submit")
print(mylink)
mylink.click()
```

Figure A-1. (*continued*)

- Figure A-2: **xls2csv.py** (a utility to mainly convert file formats from Excel to csv)

```
import re

###########################################################################
#
# Convert a downloaded ATADS file from xls format to CSV
#
# Non-data blocks are skipped, then the data rows are processed
#
# Unwanted commas are removed, and an unwanted space removed so
# airport 3 letter ids are recovered
#
###########################################################################


class my_xls2csv:
    def __init__(self):
        self.f_xls          = ''
        self.f_csv          = ''
        self.in_data_table  = False
        self.year_list      = [2014]
        self.year           = 2014
        self.res            = ''
###########################################################################
#
# Skip from top to the end of the thead block
#
def skip_through_end_of_thead_block(self):
    while (True):
        line = self.f_xls.readline()
        if '/thead>' in line:
            break

    print("leading headers skipped")
    return
```

Figure A-2. *Converting the Excel file into CSV: xls2csv.py*

191

```
###########################################################################
#
# Process a row of the XL table which is multi-line of the form
#
# <tr>
# <td> </td>
# ...
# </tr>
#
# Clean data by removing unwanted commas and fixing Airport ID field
# by removing extra space  ('ABC ' -> 'ABC')
#
# and by erasing <*> entries
#

def read_XL_row(self):
    do_read = True
    str0 = ''
    while (do_read):
        line = self.f_xls.readline()
        line = line.strip()
        if 'table_footer_lead' in line:                # End of data table
            self.in_data_table=False

        elif "<tr" in line:                                    # XL row start
            str0 = ""

        elif "<td " in line:                           # cell start
            line0 = re.sub(',','',line)            # remove commas

                                                            # change 'ABC ' to 'ABC'
            line1 = re.sub(r'([A-Z]{3})([ ])',r'\1', line0, count=1)

                                                     #remove <> tags
            str0  = str0 + ',' + re.sub('<.*?>','',line1)

        elif "</tr" in line:                                   # End of XL row found
            do_read = False

    self.row str = str0[1:]+'\n'                # also cleanup leading ','
```

Figure A-2. (*continued*)

```
##########################################################################
#
# Convert one year of ATADS xls data
#

def do_atads_year(self):

    print ('Doing year: ',str(self.year))
    self.xls_file = 'ATADS_XLS/atads' + str(self.year) + '.xls'
    self.csv_file = 'ATADS_CSV/atads' + str(self.year) + '.csv'

    print ("Converting ", self.xls_file, " to ", self.csv_file)

    self.f_xls = open(self.xls_file,'r')
    self.f_csv = open(self.csv_file,'a')

    self.skip_through_end_of_thead_block()

    print("reading data table...\n")
    self.in_data_table=True                          # Table Start

    while self.in_data_table:
        self.read_XL_row()

        if self.in_data_table:
            self.f_csv.write(self.row_str)
        else:
            break

    self.f_csv.close()
    self.f_xls.close()

x2c = my_xls2csv()

for year in x2c.year_list:
    x2c.year = year

    x2c.do_atads_year()
```

Figure A-2. (*continued*)

- Figure A-3: **split_by_apt.py** (a utility to split the annual csv data into individual airport csv files)

```
import pandas as pd
import os
from os.path import isfile, join

#
# For each year's csv file, and for each apt, apppend the apt's data to the APT/apt
# version

names=[
                "date",         "facility",     "state",    "region",       "ddso",
                "class",        "ifri_carrier", "ifri_taxi","ifri_general","ifri_mil",
                "ifri_total",   "vfri_carrier", "vfri_taxi","vfri_general","vfri_mil",
                "vfri_total",   "i_carrier",    "i_taxi",   "i_general",    "i_mil",
                "i_total",      "loc_civ",      "loc_mil",  "loc_total",    "total_ops"
                    ]

def get_csv_year_files():
        # The path for listing items
        path = './ATADS_CSV/'

        # The list of items
        csv_files = os.listdir(path)
        return(csv_files)

year_files=get_csv_year_files()
```

Figure A-3.   *Building the airport csv files: split_by_airport.py*

```
for yrf in year_files:

    print (yrf)

    df = pd.read_csv(open('./ATADS_CSV/'+yrf))
    df.columns = names

    apt_list = df[df.columns[1]].unique()

    for apt in apt_list:

        apt_csv = './APT_CSV/'+apt+'.csv'
        dfa = df.loc[df['facility'] == apt]
        dfa.to_csv(apt_csv,mode='a',index=False, header=False)
```

Figure A-3. (*continued*)

atads_scrape.py uses **chromedriver.exe** to control Google Chrome in order to navigate the remote website. Both Chrome and **chromedriver. exe** need to be installed/present – I simply place chromedriver.exe in the ATADS_DATA_UTILS folder. At this time (early 2024), there appears to be a bug in **chromedriver.exe**, and it ignores the user-specified default download directory; the downloaded file will probably appear in Chrome's normal download area.

Notes

The strategy used is to accumulate annual xls files in ATADS_XLS. This can be very time consuming, but once done, only the current year's data needs to be downloaded. While only the current year's csv file in ATADS_CSV needs to be refreshed, **xls2csv.py** rebuilds all annual csv files. While this is inefficient, it does keep the process simple. All files in APT_CSV are rebuilt by **split_by_apt.py** – again, we choose simplicity over efficiency.

Data Update Process

[Using year 2023 as an example]

1. Set the download year in **atads_scrape.py** to "2023"

2. Use **atads_scrape.py** to download 2023's data

3. Find the downloaded file (WEB-Report-*****.xls) and rename as atads2023.xls

4. Add atads2023.xls to the ATADS_XLS folder – overwrite already existing if needed

5. Rebuild the annual csv files in ATADS_CSV from those in ATADS_XLS by running **xls2csv.py**

6. Build the individual airport csv files using **split_by_apt.py**

Directory APT_CSV should now have the updated airport files and should be copied to the atads active area.

This data management scheme could be easily modified for greater efficiency and automated to run monthly using a Unix **cron** job if desired.

Index

A

add_titles() function, 30

Airshow, 166, 168

Air Traffic Activity System (ATADS)
        data, 39, 53
    Airport operators report, 40
    convert Excel to CSV with data
        cleanup, 48, 49
    output, 42
    plain text version,
        downloaded file, 44
    screen-scraping program, 45–48
    segment, downloaded file, 45
    Split_by_apt.py, 50
    web page, 40

APT_CSV, 49, 50, 185

ATADS_CSV, 185

ATADS dashboard, 90, 91, 142
    app.layout(), 90, 92
    atads_figures class, 113
    banner, 92, 93, 95
    charts, 113
    code testing, 113
    CSS, 114
    enhancements, 89, 90
    histogram panels, monthly/
        weekday, 96

add_black_border()
    method, 96
    atads_layout class, 98, 99
    className, 98, 99
    CSS file, 99, 100
    ticktext/tickvals, 97, 98
    update_mchart()
        method, 96, 97
    update_wchart()
        method, 97, 98
improvements, 114
instruction panels, 92, 93, 95
names, 113
spectrum panel, 100, 101
    amplitudes/frequencies, 103
    atads_layout(), 101
    creation, 102
    CSS file, 101
    Discrete Fourier
        Transform, 102
    fft() function, 103
    time units, 103
    update_spectrum() method,
        102, 104
    x-axis/y-axis, 102
update_dashboard(), 90, 92
weekly/seasonal effects

© Padraig Houlahan 2024
P. Houlahan, *Prototyping Python Dashboards for Scientists and Engineers*,
https://doi.org/10.1007/979-8-8688-0221-8

B

C

Printed in the United States
by Baker & Taylor Publisher Services